Math Mammoth
Grade 3 Review Workbook

By Maria Miller

Math Mammoth Grade 3 Review Workbook
Contents

Introduction

Math Mammoth Grade 3 Review Workbook is intended to give students a thorough review of third grade math, following the main areas of Common Core Standards for grade 3 mathematics. The book has both topical as well as mixed (spiral) review worksheets, and includes both topical tests and a comprehensive end-of-the-year test. The tests can also be used as review worksheets, instead of tests.

You can use this workbook for various purposes: for summer math practice, to keep the child from forgetting math skills during other break times, to prepare students who are going into fourth grade, or to give third grade students extra practice during the school year.

The topics reviewed in this workbook are:

- addition and subtraction
- multiplication concept and tables
- clock
- Money
- place value with thousands
- geometry
- measuring
- division
- fractions

In addition to the topical reviews and tests, the workbook also contains many cumulative (spiral) review pages.

The content for these is taken from *Math Mammoth Grade 3 Complete Curriculum*, so naturally this workbook works especially well to prepare students for grade 4 in Math Mammoth. However, the content follows a typical study for grade 3, so this workbook can be used no matter which math curriculum you follow.

Please note this book does not contain lessons or instruction for the topics. It is not intended for initial teaching. It also will not work if the student needs to completely re-study these topics (the student has not learned the topics at all). For that purpose, please consider *Math Mammoth Grade 3 Complete Curriculum*, which has all the necessary instruction and lessons.

I wish you success with teaching math!

Maria Miller, the author

Addition and Subtraction Review

1. Solve in your head.

a. 303 + 5 = _____	**b.** 160 + 70 = _____	**c.** 998 − 4 = _____
299 + 5 = _____	459 + 6 = _____	202 − 4 = _____

2. Write the Roman numerals using normal numbers, and the numbers using Roman numerals.

a. VI	**b.** LVI	**c.** LXV	**d.** XLVIII
e. 8	**f.** 14	**g.** 23	**h.** 67

3. Subtract.

a.
```
   4 0 5
 − 2 6 6
```

b.
```
   5 1 0
 − 2 1 6
```

c.
```
   8 0 7
 − 4 2 9
```

d.
```
   5 0 3
 − 1 2 6
```

e.
```
   4 1 5
 − 2 4 9
```

4. Write an addition and a subtraction sentence using the given numbers.

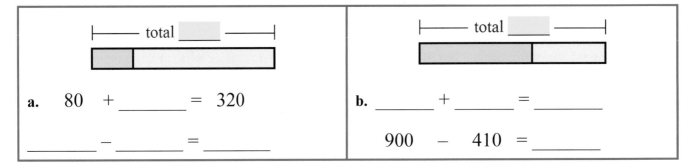

a. 80 + _____ = 320

_____ − _____ = _____

b. _____ + _____ = _____

900 − 410 = _____

5. Solve in your head.

Add up to find the difference:	**b.** 63 − 27 = _____
a. 71 − 26 = _____ + ☐ + ☐ + ☐ 26 30 70 71	**c.** 82 − 51 = _____
	d. 91 − 86 = _____

6. Calculate.

a. $50 - 20 - 5 + 6 =$ _____	**c.** $(500 - 50) + (70 - 10) =$ _____
b. $50 - (20 - 5) + 6 =$ _____	**d.** $500 - (50 + 70 - 10) =$ _____

7. Round these numbers to the nearest ten.

a. $12 \approx$ _____	**b.** $677 \approx$ _____	**c.** $46 \approx$ _____

8. Solve.

a. $693 - (800 - 134) =$ _____

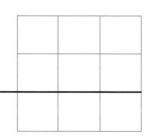 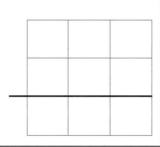

b. _____ $- 318 = 467$

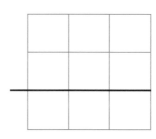

9. Solve the word problems.

a. There are 800 beads in a bag. Some are yellow, some are red, and some are blue. If there are 270 red and 270 blue beads, find how many yellow beads are in the bag.

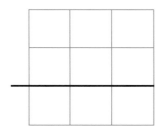 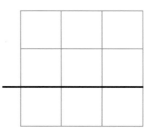

b. A store sells CDs in boxes of 100. Ann bought three full boxes and one box from which 14 CDs had been sold earlier. How many CDs did she buy?

Addition and Subtraction Test

1. Solve in your head and then write the answers.

a. $210 + 60 =$ _____	**b.** $55 + 38 =$ _____	**c.** $82 - 35 =$ _____
$198 + 5 =$ _____	$99 + 30 =$ _____	$880 - 9 =$ _____

2. Solve what number goes in place of the triangle.

a. $52 - \triangle = 47$ $\triangle =$ _____	**b.** $\triangle - 20 = 267$ $\triangle =$ _____	**c.** $693 + \triangle = 701$ $\triangle =$ _____

3. Write the Roman numerals using normal numbers.

a. IV	**b.** LXVI	**c.** LXXVIII
d. CXLIV	**e.** XXIX	**f.** XCVIII

4. Jamie has $250. He bought a camera for $127 and batteries for $18. How much money does he have left?

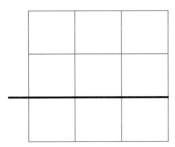 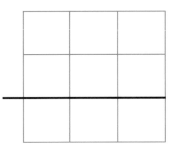

5. Subtract. *Check* the result of each subtraction by adding.

		Check:			Check:
a.	4 0 4 − 1 5 7	+	**b.**	7 2 3 − 3 9 7	+

8

6. Round the numbers to the nearest ten.

| a. 708 ≈ _____ | b. 595 ≈ _____ | c. 824 ≈ _____ | d. 457 ≈ _____ |

7. Calculate.

| a. $70 - 40 - 8 + 5 =$ _____ | c. $(300 - 30) + (60 - 20) =$ _____ |
| b. $70 - (40 - 8) + 5 =$ _____ | d. $300 - 30 + (70 - 20) =$ _____ |

8. One year has 365 days. Of those, 206 are school days.
   How many days in a year are not school days?

9. Jason has four boxes of trading cards. Each
   of the boxes contains 80 cards, except
   the last box has 28 cards missing.
   How many trading cards does he have?

10. Fill in the
    missing parts.

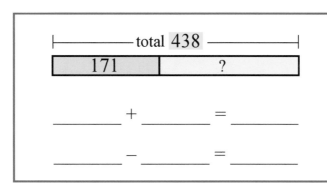

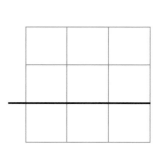

11. Solve.

$609 - (169 + 145) =$ _____

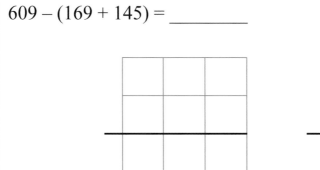

9

Concept of Multiplication Review

1. Draw a picture to illustrate the multiplications.

a. 4×5	**b.** 3×6

2. Write each multiplication as an addition.

    a. $3 \times 7 = $ _____

    b. $4 \times 20 = $ _____

3. Write the multiplication sentence.

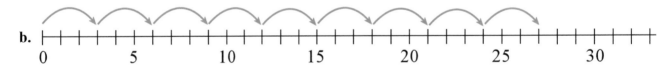

a.

    _____ × _____ = _____

b.

    _____ × _____ = _____

4. Multiply.

a. $2 \times 2 = $ _____ $1 \times 4 = $ _____	**b.** $2 \times 10 = $ _____ $3 \times 3 = $ _____	**c.** $12 \times 0 = $ _____ $12 \times 1 = $ _____
d. $0 \times 5 = $ _____ $2 \times 7 = $ _____	**e.** $2 \times 40 = $ _____ $3 \times 30 = $ _____	**f.** $2 \times 400 = $ _____ $1 \times 500 = $ _____

5. Solve.

a. Each bag holds five balls. How many balls are in four bags?
b. Write a multiplication to show how many legs five horses have.
c. The cat eats two cans of cat food each day. Last week it was sick and didn't eat any food for two days. How many cans of cat food did it eat during that week?
d. During one week, Karen read three books each day, except on Thursday, when she read only one book. How many books did she read in total that week?

6. Calculate in the correct order.

a. $(3 + 2) \times 2$	**b.** $2 \times 10 - 1 \times 3$
c. $12 + 3 \times 5 - 4$	**d.** $(1 + 7) \times 2 - 4$

7. Skip-count and fill in **the multiplication table of 4**. Draw arrows on the number line to illustrate the skip counting.

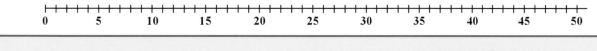

$1 \times 4 = $ _____ $4 \times 4 = $ _____ $7 \times 4 = $ _____ $10 \times 4 = $ _____

$2 \times 4 = $ _____ $5 \times 4 = $ _____ $8 \times 4 = $ _____ $11 \times 4 = $ _____

$3 \times 4 = $ _____ $6 \times 4 = $ _____ $9 \times 4 = $ _____ $12 \times 4 = $ _____

Concept of Multiplication Test

1. Multiply.

a. $2 \times 3 =$ _____	**b.** $2 \times 5 =$ _____	**c.** $2 \times 20 =$ _____	**d.** $1 \times 9 =$ _____
$1 \times 5 =$ _____	$3 \times 10 =$ _____	$3 \times 40 =$ _____	$11 \times 0 =$ _____
$0 \times 7 =$ _____	$2 \times 6 =$ _____	$2 \times 200 =$ _____	$11 \times 1 =$ _____

2. Draw a picture to illustrate the problems.

a. 3×5	**b.** $2 \times 5 + 3 \times 4$

3. Write a number sentence for each problem and solve.

a. Each basket holds 12 apples.
How many apples are in three baskets?

b. Chloe bought four pens for $2 each and two games for $8 each.
What was the total bill?

c. You have 20 sticks, and you make groups of four sticks.
How many groups can you make?

4. Calculate.

a. $5 + 3 \times 5$	**b.** $20 + 2 \times 3 - 4$
c. $0 \times (10 + 2) \times 3$	**d.** $(8 - 3) \times 1 + 6$

Mixed Review 1

1. Write the Roman numerals using normal numbers.

a.	b.	c.	d.
III	XV	IX	XXIV
VII	XXIII	LXI	CLXXV

2. Add mentally.

a. $93 + 6 =$ _____	**b.** $47 + 29 =$ _____	**c.** $15 + 18 =$ _____
$893 + 6 =$ _____	$607 + 9 =$ _____	$624 + 8 =$ _____

3. Subtract in parts: First, subtract to the previous whole ten, then subtract the rest.

a. $161 - $ _6_	**b.** $332 - 5$	**c.** $773 - 8$
$161 - $ _1_ $ - $ _5_	$332 - $ ____ $ - $ ____	$773 - $ ____ $ - $ ____
$=$ _____	$=$ _____	$=$ _____

4. Fill in the missing parts so that the addition and subtraction sentences match the model.

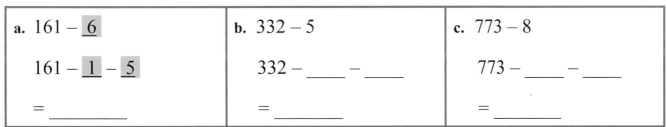

a. $240 + $ _____ $= 400$	**b.** $360 + $ _____ $= 410$
_____ $-$ _____ $=$ _____	_____ $-$ _____ $=$ _____

5. Calculate.

a. $19 - (6 + 2) + 5 =$ _____	**b.** $(800 - 60) - (50 - 40) =$ _____
$19 - 6 + 2 + 5 =$ _____	$800 - 60 - 50 - 40 =$ _____

6. Subtract. Be careful with regrouping! Check by adding.

a.	b.
8 3 5 − 5 7 6 + _____	6 0 2 − 4 2 6 + _____

7. Solve the word problems.

a. Danny ran three times around a track that was 245 yards long. How long a distance did he run?

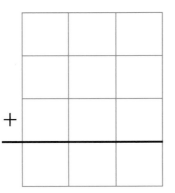

b. It is 350 km from Jon's home to his grandpa's place. The family drives 176 km of that distance and stops for a rest. How long do they still have to go?

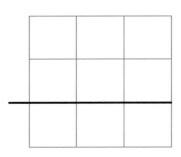

c. One jar had 315 beans and another had 50 fewer beans than that. How many beans are in the <u>two</u> jars?

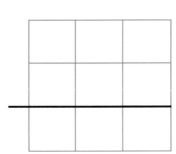 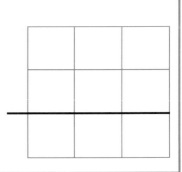

8. Find *about* how much the two things cost together. Use rounded numbers!

a.	b.
a toy, $28, and a set of books, $129	a ladder, $62, and wheelbarrow, $137
toy about $_____	ladder about $_____
a set of books about $_____	wheelbarrow about $_____
together about $_____	together about $_____

Mixed Review 2

1. Add and subtract in your head.

a. $566 + 8 =$	**b.** $730 + 80 =$	**c.** $991 - 8 =$

2. Write the numbers as Roman numerals.

    a. 25   **b.** 19   **c.** 57   **d.** 143

3. Calculate.

a. $35 - 14 - 7 + 3 =$ _____	**d.** $(250 - 20) + (80 - 30) =$ _____
b. $35 - (14 - 7) + 3 =$ _____	**e.** $250 - (20 + 80 - 30) =$ _____
c. $35 - (14 - 8 + 3) =$ _____	**f.** $250 - 20 + (80 - 30) =$ _____

4. Jill has a tea set with 14 cups and another tea set with 13 cups.
   She wants to invite the girls from her class for a tea party. There
   are 30 girls in her class. How many more cups does she need?

5. Multiply.

a. $3 \times 3 =$ ____	**b.** $5 \times 2 =$ ____	**c.** $3 \times 30 =$ ____	**d.** $0 \times 9 =$ ____
$4 \times 2 =$ ____	$3 \times 6 =$ ____	$4 \times 20 =$ ____	$10 \times 0 =$ ____
$0 \times 8 =$ ____	$4 \times 8 =$ ____	$2 \times 400 =$ ____	$22 \times 1 =$ ____

6. Write the multiplications as additions, and solve.

    a. 2×20

    b. 3×50

7. Write a number sentence or sentences for each problem and solve them.

a. Tim has three rolls of string with 12 feet
on each roll. What is the total length of
the string on the three rolls?

b. Julie has 16 golf balls and 8 tennis balls.
She put the balls into bags with four in each bag.
How many bags does she need?

8. Write one addition and one subtraction sentence to match the model.

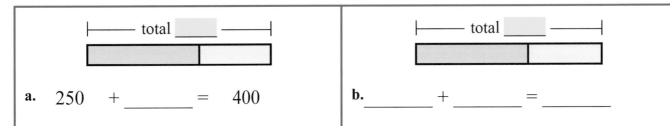

a.   250   + _____ =   400

_____ − _____ = _____

b. _____ + _____ = _____

500   − _____ =   390

9. Subtract.

a.   8 8 8   **b.**   4 5 0   **c.**   6 0 2   **d.**   8 0 0
     − 2 9 9   − 1 3 4   − 3 4 4   − 6 5 7

10. Can you buy three bicycles for $48 each and pay with $150?

If yes, how much money will you have left?

If no, how much more money would you need?

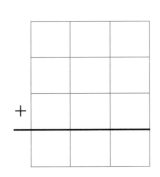

11. Estimate the total cost using rounded numbers.

a. Rent, $256. Groceries, $387.	**b.** Ticket for adult, $58. Ticket for child, $38.
Rent: about $_____	Ticket for adult: about $_____
Groceries: about $_____	Ticket for child: about $_____
Total: about $_____	Total cost: about $_____

Multiplication Tables Review

1. Fill in the table—for the last time.

×	0	1	2	3	4	5	6	7	8	9	10	11	12
0													
1													
2													
3													
4													
5													
6													
7													
8													
9													
10													
11													
12													

2. Compare the expressions and write $<$, $>$, or $=$.

a. $9 \times 8 \boxed{} 10 \times 8$

b. $9 \times 5 \boxed{} 11 \times 4$

c. $9 \times 2 \boxed{} 3 \times 6$

d. $9 \times 8 \boxed{} 9 \times 4$

e. $4 \times 4 \boxed{} 2 \times 8$

f. $10 \times 11 \boxed{} 10 \times 7$

g. $10 \times 8 \boxed{} 10 \times 5$

h. $9 \times 2 \boxed{} 4 \times 5$

i. $9 \times 8 \boxed{} 9 \times 6$

3. Fill in the tables:

$1 \times 3 =$ _____	$7 \times 3 =$ _____	$1 \times 6 =$ _____	$7 \times 6 =$ _____
$2 \times 3 =$ _____	$8 \times 3 =$ _____	$2 \times 6 =$ _____	$8 \times 6 =$ _____
$3 \times 3 =$ _____	$9 \times 3 =$ _____	$3 \times 6 =$ _____	$9 \times 6 =$ _____
$4 \times 3 =$ _____	$10 \times 3 =$ _____	$4 \times 6 =$ _____	$10 \times 6 =$ _____
$5 \times 3 =$ _____	$11 \times 3 =$ _____	$5 \times 6 =$ _____	$11 \times 6 =$ _____
$6 \times 3 =$ _____	$12 \times 3 =$ _____	$6 \times 6 =$ _____	$12 \times 6 =$ _____

In the tables above, color the number (the answer) orange, if it is in both tables.
What do you notice?

4. Solve the problems.

a. The class has eleven girls. They each have seven schoolbooks.
How many schoolbooks do the girls have in total?

b. The teacher puts 20 students in groups so that each group has 4 students.
How many groups will there be?

c. Josefina bought four books of stickers that cost $3 each and a notebook for $7.
What was the total cost?

d. Andy bought some packages of seeds for $24. Each package cost $2.
How many packages did he buy?

e. A zoo has five s, three s, and twenty s.
How many feet do those animals have all totaled?

5. Find the missing factors.

a.	b.	c.	d.
____ × 8 = 24	6 × ____ = 18	7 × ____ = 49	____ × 5 = 25
____ × 8 = 64	6 × ____ = 66	____ × 7 = 56	____ × 5 = 45
____ × 8 = 40	6 × ____ = 12	7 × ____ = 63	____ × 5 = 35

e.	f.	g.	h.
____ × 4 = 16	____ × 3 = 36	____ × 8 = 48	____ × 12 = 60
____ × 4 = 28	____ × 3 = 21	____ × 8 = 32	____ × 12 = 84
4 × ____ = 36	3 × ____ = 27	8 × ____ = 72	12 × ____ = 108

Mystery Number

(All mystery numbers are less than 100.)

a. You can find me both in the table of eleven and in the table of four.

I am _____.

b. I am more than 15. I am in the table of two, the table of three, and the table of four!

I am _____.

c. I am between 15 and 35. The number one more than me is in the table of five. The number one less than me is in the table of four.

I am _____.

d. I am both in the table of four and in the table of three, and if you add one to me, I am in the table of five.

I am _____.

e. I am in the table of 11. The number that is one more than me, is in the table of five, but not in the table of ten.

I am _____.

f. I am less than 22 but more than 9, and I am in the table of four. If you exchange my digits, I am in the table of three!

I am _____.

19

Multiplication Tables Test

1. Fill in the complete multiplication table!

×	0	1	2	3	4	5	6	7	8	9	10	11	12
0													
1													
2													
3													
4													
5													
6													
7													
8													
9													
10													
11													
12													

2. **a.** Which multiplication fact is both in the table of 3 and in the table of 8?

   b. Which multiplication fact is both in the table of 9 and in the table of 7?

3. Solve the problems.

a. A pet store has 10 kittens for sale. Five of them cost $9 each and the rest cost $5 each. How much would all 10 kittens cost?

b. If one table can seat six people, how many tables do you need for 54 people who are coming to the restaurant?

c. Ann saw seven dogs, four cats, and twelve geese at the park. How many feet do the animals have in total?

d. A T-shirt costs $6. How many shirts can you buy with $48?

4. Find the missing factors.

a.	b.	c.	d.
_____ × 6 = 24	7 × _____ = 77	5 × _____ = 35	_____ × 3 = 27
_____ × 6 = 54	7 × _____ = 42	_____ × 5 = 20	_____ × 3 = 12
_____ × 6 = 36	7 × _____ = 14	5 × _____ = 55	_____ × 3 = 36

e.	f.	g.	h.
_____ × 11 = 66	_____ × 8 = 64	_____ × 4 = 24	_____ × 12 = 144
_____ × 11 = 121	_____ × 8 = 16	_____ × 4 = 36	_____ × 12 = 48
11 × _____ = 22	8 × _____ = 32	4 × _____ = 16	12 × _____ = 84

Mixed Review 3

1. Find the difference of:

a. 24 and 51	b. 300 and 987	c. 437 and 442

2. Round the numbers to the nearest ten.

a. 663 ≈ _____	b. 598 ≈ _____	c. 815 ≈ _____	d. 64 ≈ _____

3. Solve. Remember the order of operations.

a. 563 + (409 − 226) = _____

b. 902 − 444 + 263 = _____

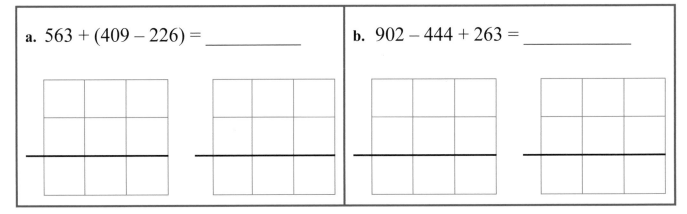

4. Solve the problem. Write an addition and a subtraction to match it.

A box has 52 candles; 16 are red and the rest are white. How many are white?

_____ + _____ = _____

_____ − _____ = _____

5. Write using Roman numerals.

a. 12	b. 34	c. 55	d. 80

6. Solve what number goes in place of the triangle.

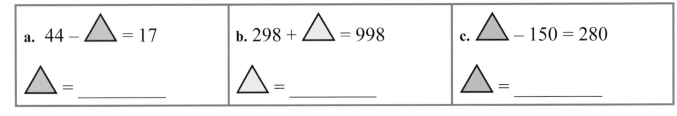

a. 44 − △ = 17

△ = _____

b. 298 + △ = 998

△ = _____

c. △ − 150 = 280

△ = _____

7. Draw dots in groups to show the multiplications.

a. 3 × 5	**b.** 4 × 3

8. Write a multiplication sentence for each problem. You can draw pictures to help.

a. How many toes do five children have in total?	_____ × _____ = _____
b. Jack arranged his 15 toy cars in rows. He put five cars in each row. How many rows did he get?	_____ × _____ = _____
c. One round table can seat four people. How many tables do you need for 20 people?	_____ × _____ = _____

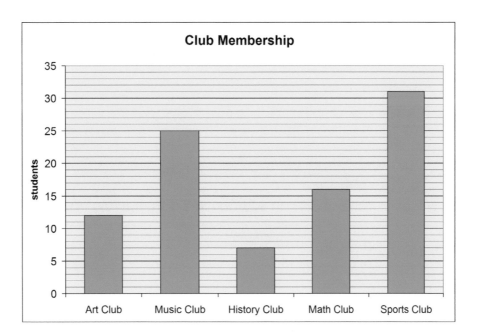

9. **a.** Which is the most popular club?

b. How many more students are in the sports club than in the math club?

c. How many students are in the art, music, and sports clubs in total?

23

Mixed Review 4

1. Jimmy rode his bike from Brigham City to American Fork in two days, riding the same distance each day.
   How many miles did he ride each day?

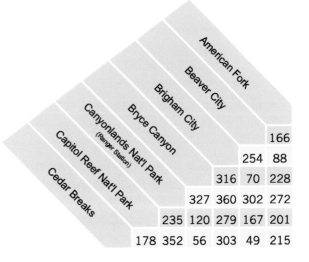

2. Ben and Joe had a three-legged journey:
   (1) They took a bus from Beaver City to Bryce Canyon.
   (2) A friend took them from there to Capitol Reef National Park.
   (3) Then they rode on a bus from there to Brigham City.

   What was the total number of miles they traveled?

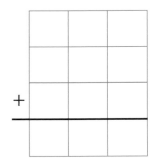

3. Three parts make up one whole. Write an addition and a subtraction sentence, and solve.

total 698 ? | 196 | 153 **a.** _____ + _____ + _____ = _____ _____ − _____ − _____ = _____	
total 450 125 | 250 | ? **b.** _____ + _____ + _____ = _____ _____ − _____ − _____ = _____	

4. Solve the problems.

a. A math teacher bought four calculators that cost $8 each and ten notebooks that cost $2 each. What is the total cost?

b. Sheila and three other girls equally shared the cost of a taxi to the mall, which was $12. At the mall, Sheila bought a book for $6. How much did Sheila spend for the taxi fare plus the book?

5. Write the Roman numerals using normal numbers.

a. IX   **b.** CXXI   **c.** LXVII   **d.** XIV

6. Write multiplication sentences for the jumps on the number lines below.

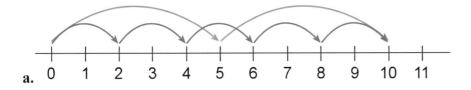

a.   0 1 2 3 4 5 6 7 8 9 10 11

_____ × _____ = _____

_____ × _____ = _____

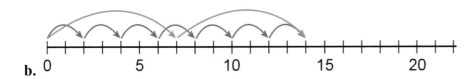

b.   0   5   10   15   20

_____ × _____ = _____

_____ × _____ = _____

7. Some scientists were studying giraffes in a park in Africa. The pictograph shows how many giraffes they saw at a waterhole each week. One 🦒 means 15 giraffes.

a. How many giraffes did they see in week 27?

b. How many more giraffes did they see in week 27 than in week 28?

Week 25	🦒🦒🦒🦒
Week 26	🦒🦒🦒🦒🦒🦒
Week 27	🦒🦒🦒🦒🦒🦒🦒
Week 28	🦒🦒🦒

Telling Time Review

1. Write the time the clock shows. Below, write the time 10 minutes later.

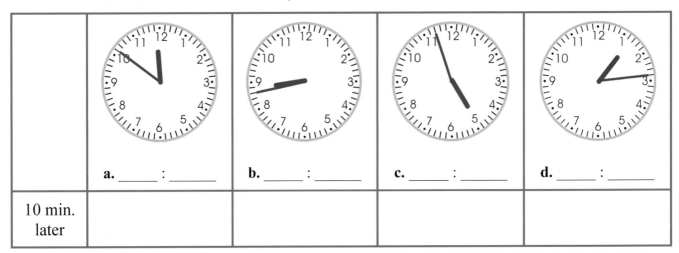

	a. _____ : _____	b. _____ : _____	c. _____ : _____	d. _____ : _____
10 min. later				

2. How many minutes is it from the time on the clock face until the given time?

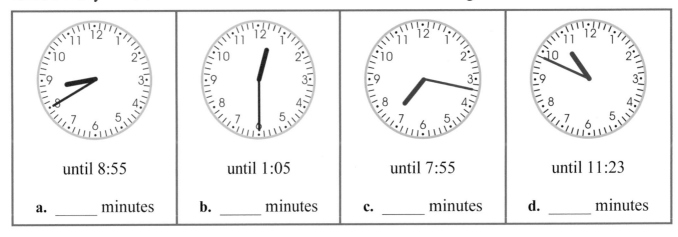

until 8:55	until 1:05	until 7:55	until 11:23
a. _____ minutes	b. _____ minutes	c. _____ minutes	d. _____ minutes

3. How much time passes between the two times given?

a. from 4:08 until 10:08	b. from 3 AM until 5 PM
c. from 8:23 until 8:41	d. from 3:37 until 4:17

4. The music class starts at 1:45 and ends 50 minutes later.
   At what time does it end?

5. The train left at 11:10 and arrived at 12:20 PM.
   How long was the trip?

Telling Time Test

1. Write the time the clock shows, and the time 10 minutes later.

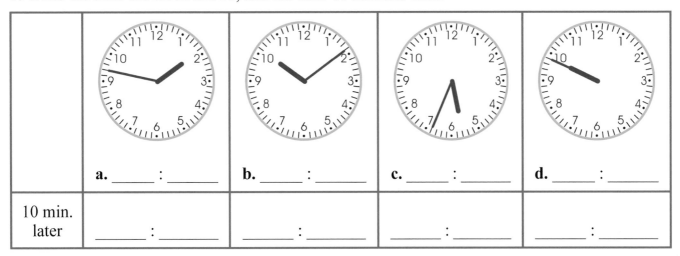

a. _____ : _____   b. _____ : _____   c. _____ : _____   d. _____ : _____

| 10 min. later | _____ : _____ | _____ : _____ | _____ : _____ | _____ : _____ |

2. How much time passes?

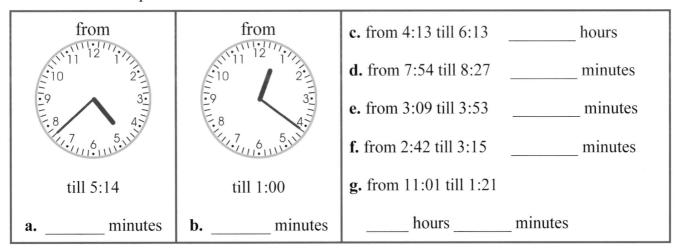

from

till 5:14

a. _____ minutes

from

till 1:00

b. _____ minutes

c. from 4:13 till 6:13 _____ hours

d. from 7:54 till 8:27 _____ minutes

e. from 3:09 till 3:53 _____ minutes

f. from 2:42 till 3:15 _____ minutes

g. from 11:01 till 1:21

_____ hours _____ minutes

3. Solve the problems.

a. The ballet starts at 7:15 and ends 75 minutes later. When does it end?

b. The plane left at 9:25 AM and arrived at 1:00 PM. How long was the trip?

c. Denny left for the orchestra practice at 6:15 PM and arrived back home at 8:30 PM. How long was he gone?

Mixed Review 5

1. Add and subtract in your head.

a. 76 − 51 = _____	**b.** 82 − 39 = _____	**c.** 65 − 36 = _____
d. 385 + 4 = _____	**e.** 552 + 9 = _____	**f.** 795 + 8 = _____

2. Write the Roman numerals using normal numbers.

  a. XIV   **b.** LXVI   **c.** XLIX   **d.** CXL

3. One MP3 player costs $89 and another costs $17 less than the first. If you buy two of the cheaper players, how much do they cost together?

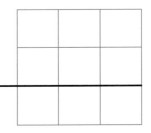

4. Multiply.

a.	**b.**	**c.**	**d.**
9 × 3 = _____	12 × 2 = _____	7 × 8 = _____	6 × 6 = _____
7 × 4 = _____	3 × 8 = _____	9 × 7 = _____	12 × 5 = _____
10 × 0 = _____	6 × 4 = _____	4 × 8 = _____	3 × 7 = _____

5. Solve the problems. Write a calculation (multiplication and/or addition and/or subtraction) for each problem. You can also draw pictures to help!

a. Every day for a week, Derek read two books. How many books did he read in total?

    _____ × _____ = _____

b. Jack put three pencils in each of the seven pencil cases, and in the eighth one he put five. How many pencils did Jack put in the pencil cases?

6. Calculate. Circle the operation to be done first. Parenthesis → multiply → add/subtract.

a. $2 + 5 \times 2$	**b.** $5 \times (1 + 1)$	**c.** $(4 - 2) \times 7$

7. Match the correct addition and/or subtraction with the problem. Then solve.

a. Jack weighs 141 pounds, and Davy is 22 lb
lighter than him. How much does Davy weigh?

$141 + 22 = \underline{?}$ $141 + \underline{?} = 22$

$22 + \underline{?} = 141$ $141 - \underline{?} = 22$

b. Liz compared the prices of two washers. One was
$48 cheaper than the other, which cost $275.
How much does the cheaper one cost?

$275 + 48 = \underline{?}$ $\underline{?} + 48 = 275$

$275 - 48 = \underline{?}$ $\underline{?} - 48 = 275$

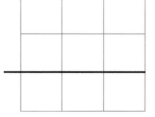

8. Estimate, using rounded numbers, the total distance all the way around this rectangular path.

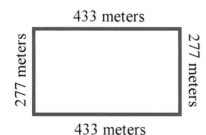

433 meters

277 meters

277 meters

433 meters

Mixed Review 6

1. For each problem, write a corresponding subtraction or addition sentence, and solve.

a. _____ + 120 = 770	b. _____ + _____ = _____
_____ − _____ = _____	_____ − 9 = 633

2. Read the list of numbers below from left to right, and solve.

$$12, \quad 14, \quad 21, \quad 33, \quad 87, \quad 32, \quad 435, \quad 54, \quad 89, \quad 100$$

   a. Add the first and second numbers.

   b. Subtract the fourth number from the ninth number.

   c. Add the eighth number to the tenth number and
   then subtract the fifth number.

3. Calculate.

a. $(18 - 5) - (3 + 6) =$ _____	b. $(300 - 50) - (80 - 30) =$ _____
$18 - 5 - 3 + 6 =$ _____	$300 - 50 - 80 - 30 =$ _____

4. Solve the problems.

   a. Karen baked 30 cupcakes. She ate one. Her brother took two. Then her mother said she
   needed 2 cupcakes for herself and each of the twelve ladies coming for afternoon tea.

   Are there enough cupcakes left?

   If not, how many more cupcakes
   does Karen need to make?

   b. The teacher gave each of the nine children 12 marbles to play a math game.
   After the class, only 99 marbles were gathered back.
   How many marbles were lost?

5. Write the time as (hours) : (minutes).

a. 8 past 6	**b.** 12 till 7	**c.** 29 past 3	**d.** 33 past 5
_____ : _____	_____ : _____	_____ : _____	_____ : _____
e. 24 till 5	**f.** 21 till 6	**g.** 2 till 12	**h.** 17 till 1
_____ : _____	_____ : _____	_____ : _____	_____ : _____

6. Calculate.

a. $8 \times 10 - 2 + 5 = $ _____	**b.** $6 + 7 \times (4 - 2) = $ _____
c. $3 \times 4 - 2 \times 3 = $ _____	**d.** $2 \times (4 + 4) \times 2 = $ _____

7. Continue the patterns.

a. $564 - 5 = $ _____	**b.** $888 + 12 = $ _____
$564 - 10 = $ _____	$886 + 14 = $ _____
$564 - 15 = $ _____	$884 + 16 = $ _____
$564 - $ ____ $ = $ _____	_____ $ + $ ____ $ = $ _____
$564 - $ ____ $ = $ _____	_____ $ + $ ____ $ = $ _____
$564 - $ ____ $ = $ _____	_____ $ + $ ____ $ = $ _____

8. Solve the problems.

a. Grace stayed at the beach for three weeks. Then she went to her grandparent's farm for two weeks. How many *days* did she spend at the beach and the farm in total?

b. Ben walks to school in fifteen minutes. From Monday through Friday, how many minutes does he spend walking to and from school?

Money Review

1. How much money? Write the amount.

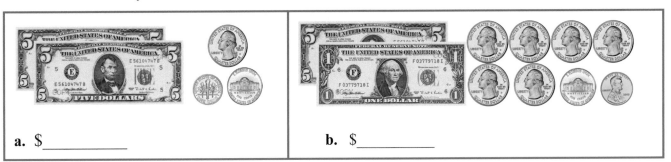

a. $_____

b. $_____

2. Write as dollar amounts.

five dimes and a quarter	three half-dollars, three nickels, and 8 pennies	three quarters, two dimes, and a half-dollar
a. $_____	b. $_____	c. $_____

3. Solve the problems.

a. Maria has $23. She wants to buy a game for $42.95. How much more money does she still need?	**b.** Arnold bought a sandwich for $2.55, soup for $2.30, and juice for $1.85. Find the total bill.	**c.** What is Arnold's change from $10?

4. Solve using mental math.

a. If you buy stickers for $2.35 and a notebook for $1.20, what is the total cost?

b. What is your change from $5?

Money Test

1. How much money? Write the amount.

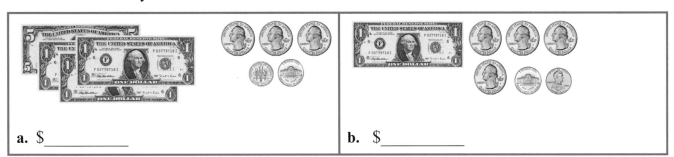

a. $_____

b. $_____

2. Write as dollar amounts.

2 nickels, 3 dimes and 4 quarters	two quarters, six dimes, and 11 pennies	three quarters, four dimes, and a penny
a. $_____	b. $_____	c. $_____

3. Solve in your head.

   a. You bought stamps for $2.25, a pen for $0.75, and a notebook for $1.30. What was the total cost?

   b. What is your change from $5?

4. Solve.

a. Marsha has saved $25. She wants to buy a game for $41.85. How much money does she still need to save?	b. Mike bought a sandwich for $3.45, soup for $2.25, juice for $1.65, and water for $1.16. Find the total cost.	c. Find Mike's change from $20.

Mixed Review 7

1. Draw jumps to fit the multiplication problem.

```
|--+--+--+--+--+--+--+--+--+--+--+--+--+--+--+--+--+--+--+--+--+--+--+--+--+--+--+--+--+--+--|
0        5        10        15        20        25        30
```

a. $7 \times 4 = $ _____

```
|--+--+--+--+--+--+--+--+--+--+--+--+--+--+--+--+--+--+--+--+--+--+--+--+--+--+--+--+--+--+--|
0        5        10        15        20        25        30
```

b. $3 \times 7 = $ _____

2. Add parentheses to each equation to make it true.

a. $50 - 20 - 7 = 37$	**b.** $8 - 5 \times 2 - 1 = 5$	**c.** $15 + 5 \times 2 - 1 = 20$

3. Multiply.

a.	**b.**	**c.**	**d.**
$9 \times 8 = $ _____	$8 \times 6 = $ _____	$6 \times 6 = $ _____	$7 \times 12 = $ _____
$7 \times 7 = $ _____	$5 \times 7 = $ _____	$9 \times 9 = $ _____	$8 \times 8 = $ _____
$9 \times 6 = $ _____	$7 \times 4 = $ _____	$12 \times 6 = $ _____	$6 \times 3 = $ _____

4. Solve the problems. Write a calculation (multiplication and/or addition and/or subtraction) for each problem. You can also draw pictures to help!

a. A pair of socks costs $5. How many pairs can you buy with $40?

_____ × _____ = _____

b. You can fit 7 dominos in one layer in their box. How many layers will there be with 28 dominoes?

5. Draw a bar model to match the addition or subtraction. Fill in the missing parts.

a. _____ + _____ = _____

998 − 500 = _____

b. 203 + _____ = 304

_____ − _____ = _____

6. Find the missing factors.

a. ____ × 7 = 21	b. 9 × ____ = 81	c. 8 × ____ = 64	d. ____ × 6 = 36
____ × 7 = 49	9 × ____ = 54	8 × ____ = 48	____ × 6 = 54
____ × 7 = 63	9 × ____ = 45	8 × ____ = 32	____ × 6 = 42

7. Solve the word problems.

a. An adult ticket to the zoo costs $54 and a child's ticket costs $34. How much do two adult tickets and one child's ticket cost in total?

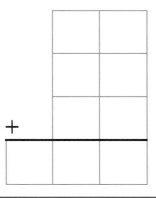

b. You are on page 265 of a book that has 317 pages. How many more pages do you still have to read?

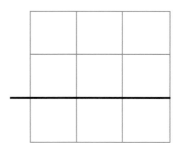

c. A bag contains 250 ribbons in three different colors. There are 89 green ones, 76 pink ones, and the rest are blue. How many are blue?

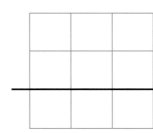

Mixed Review 8

1. Add in your head.

a. $49 + 13 =$ _____	**b.** $46 + 15 =$ _____	**c.** $25 + 39 =$ _____

2. Break the second number into its tens and ones, then subtract the parts one at a time.

a.	**b.**	**c.**
$98 - \boxed{66}$	$54 - \boxed{26}$	$73 - \boxed{17}$
$98 - \underline{60} - \underline{6} =$ _____	$54 - \underline{} - \underline{} =$ _____	$73 - \underline{} - \underline{} =$ _____

3. Write $<$, $>$ or $=$.

$\quad\quad$ **a.** $350 - 18 \ \boxed{}\ 350 - 15$ $\qquad\qquad\qquad$ **b.** $180 - 15 \ \boxed{}\ 190 - 25$

$\quad\quad$ **c.** $264 + 7 \ \boxed{}\ 267 + 8$ $\qquad\qquad\qquad$ **d.** $62 - 27 \ \boxed{}\ 61 - 27$

4. Potatoes cost \$0.32 a pound. You buy three pounds
   and pay with a \$5-bill. What is your change?

5. Miriam spent three weeks in Florida, two weeks in
   Maine, and nineteen days in New Jersey. How many
   days did she spend altogether in the three states?

6. Subtract. Check your answers.

a. $\quad\begin{array}{r} 9\ 0\ 4 \\ -\ 3\ 2\ 7 \\ \hline \end{array}$ $\quad +$ _____	**b.** $\quad\begin{array}{r} 8\ 1\ 2 \\ -\ 3\ 2\ 7 \\ \hline \end{array}$ $\quad +$ _____

7. Sally made fridge magnets and sold them at the flea market last Saturday.

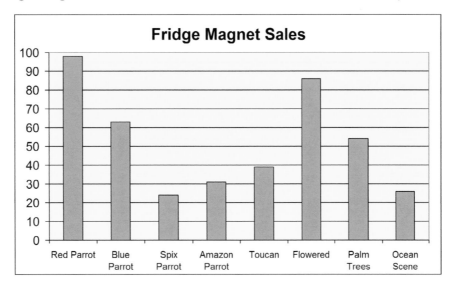

a. *About* how many parrot magnets did she sell in total
(red parrot, blue parrot, Spix parrot, and Amazon parrot magnets)?

b. *About* how many other kinds of magnets did she sell?

8. Calculate in the correct order.

a. $7 + 4 \times 7$	b. $30 + 6 \times (6 - 6)$	c. $2 \times 44 - 8 \times 0$

9. How much time passes?

a. From 2:26 till 10:18	b. From 8:29 till 12:02	c. From 2:56 till 5:34

10. Write as dollar amounts.

four nickels and a dime	three quarters, one dime, and 6 pennies	two nickels, two dimes, and two quarters
a. $_____	b. $_____	c. $_____

Place Value with Thousands Review

1. Fill in the table.

a. seven thousand two hundred forty	**b.** Six thousand five	**c.** Two thousand twenty-nine
T H T O	T H T O	T H T O

2. These numbers are written as sums. Write them in the normal way.

a. $7000 + 500 + 3 = $ _____

    $3000 + 90 = $ _____

b. $30 + 1000 + 7 = $ _____

    $400 + 6000 = $ _____

3. Compare. Write $<$, $>$, or $=$ in the box.

a. $7000 + 50 \boxed{} 5000 + 7$

b. $500 + 4 + 6000 \boxed{} 6000 + 400 + 5$

c. $80 + 3000 \boxed{} 3000 + 200$

d. $\quad 400 + 80 \boxed{} 8000 + 40$

4. Add and subtract mentally.

a. $1{,}200 + 700 = $ _____

    $400 + 6{,}800 = $ _____

b. $3{,}600 - 300 = $ _____

    $4{,}200 - 500 = $ _____

c. $7{,}200 + $ _____ $= 8{,}000$

    $8{,}000 - $ _____ $= 7{,}100$

d. $3{,}400 + 1{,}500 = $ _____

    $7{,}500 + 800 = $ _____

5. Solve (find the number that the symbol stands for).

a. $3{,}400 + \triangle = 4{,}100$

$\triangle = $ _____

b. $\triangle - 600 = 9{,}200$

$\triangle = $ _____

c. $10{,}000 - \triangle = 8{,}500$

$\triangle = $ _____

6. Round these numbers to the nearest hundred.

a.	b.	c.	d.
872 ≈ _____	5,253 ≈ _____	6,034 ≈ _____	2,739 ≈ _____

7. Add and subtract. Estimate first by rounding the numbers to the nearest hundred.

a. Estimate:

   2,540   +   1,803

    ↓   ↓

     +   = _____

Calculate exactly:

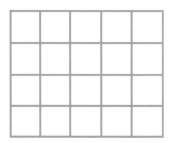

b. Estimate:

   6,581   –   736

    ↓   ↓

     –   = _____

Calculate exactly:

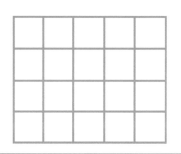

8. Solve the word problems.

a. Dad bought wood for construction for $1,616, paint for $278, and other materials for $969. Find his total bill.

Also, estimate the answer using rounded numbers.

My estimate: _____

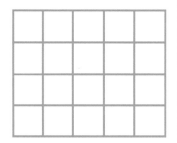

b. You have $5,000 to spend. First, you buy a pump for $278 and then some cement for $1,250. How much do you have left after that?

Also, estimate the answer using rounded numbers.

My estimate: _____

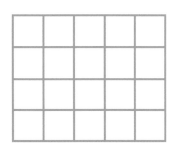

Place Value with Thousands Test

1. Write as normal numbers.

a. $2{,}000 + 600 + 80 + 9 =$ _____	**b.** $70 + 4{,}000 =$ _____
c. $600 + 9 + 5{,}000 =$ _____	**d.** $3{,}000 + 2 + 900 =$ _____

2. Compare, and write $<$, $>$ or $=$.

   a. $600 + 40$ ☐ $400 + 60 + 1$ **b.** $200 + 7{,}000$ ☐ $5{,}000 + 800$

   c. $700 + 5{,}000$ ☐ $50 + 7{,}000$ **d.** $900 + 8$ ☐ $8{,}000 + 9$

3. Add and subtract in your head and write the answers.

a. $6{,}300 +$ _____ $= 7{,}000$ $9{,}700 - 1{,}500 =$ _____	**b.** $5{,}400 + 2{,}700 =$ _____ $9{,}000 - 900 =$ _____

4. Round the numbers to the nearest hundred.

   a. $528 \approx$ _____   **b.** $1{,}384 \approx$ _____   **c.** $2{,}948 \approx$ _____

5. Add and subtract. <u>First</u> estimate by rounding.

a. Estimate: $2{,}865 \quad + \quad 4{,}531$ $\downarrow \qquad\qquad \downarrow$ $\quad + \quad\quad = $ _____	**Calculate exactly:**
b. Estimate: $7{,}002 \quad - \quad 2{,}973$ $\downarrow \qquad\qquad \downarrow$ $\quad + \quad\quad = $ _____	**Calculate exactly:**

6. Solve.

a. An animal park buys animal feed for $1,589 and tools for $325. They pay with $2,000. What is their change?

Also, estimate the answer using rounded numbers.
My estimate:

_____

b. A new computer costs $2,566 and a used one $650. What is the price difference?

Also, estimate the answer using rounded numbers.

My estimate: _____

Mixed Review 9

1. Subtract and compare the problems.

a. $56 - 7 =$ _____	b. $72 - 9 =$ _____	c. $83 - 8 =$ _____
$256 - 7 =$ _____	$672 - 9 =$ _____	$283 - 8 =$ _____

2. Write a number sentence for these word problems, and solve them.

a. How many legs do six chickens and four dogs have in total?

_____

b. Nine classrooms in the school have four windows each, and one has only three. How many windows are there in total?

_____

3. Round the numbers to the nearest ten.

a. 574 ≈ _____	b. 895 ≈ _____	c. 604 ≈ _____	d. 56 ≈ _____
342 ≈ _____	255 ≈ _____	427 ≈ _____	998 ≈ _____

4. Write the Roman numerals using normal numbers.

a. III	b. XIX	c. LXXXV	d. XLIII
VIII	XXIV	LIII	CXXV
XIV	LX	XL	CCLXXI

5. Write using Roman numerals.

a. 15	b. 21	c. 56	d. 90
19	43	65	99

6. Find the missing factors.

a. _____ × 6 = 24	b. 7 × _____ = 49	c. 8 × _____ = 64	d. _____ × 9 = 36
_____ × 6 = 54	7 × _____ = 35	8 × _____ = 48	_____ × 9 = 72
_____ × 6 = 42	7 × _____ = 56	8 × _____ = 32	_____ × 9 = 45

7. How much time passes?

a. From 5:00 AM to 10:20 AM	b. From 8:30 AM to 1:00 PM
c. From 7 PM to 6 AM	d. From 10:55 AM to 3:55 PM

8. Find the total cost of buying the items listed. Line up the numbers carefully for adding.

$2.90	$1.45	$7.50	$18.49	$6.32	$1.50

a. scissors and crayons	b. pencils, a calculator, and a book	c. two books and two calculators

Mixed Review 10

1. Find the missing factors.

a. _____ × 4 = 16	b. _____ × 9 = 0	c. _____ × 6 = 36	d. _____ × 5 = 45
_____ × 8 = 64	_____ × 3 = 27	_____ × 4 = 36	_____ × 2 = 18

2. Write using Roman numerals.

a. 15	b. 32	c. 47	d. 56

3. Write the time using the expressions "till" and "past."

    a. 6:38   **b.** 3:56

    c. 2:12   **d.** 7:43

4. Sharon started doing an exercise video at 7:35 PM, and it ended 43 minutes later. At what time did it end?

5. Estimate these calculations by rounding the numbers to the nearest hundred. On the right, calculate the exact answer.

a. **Estimate:**	**Calculate exactly:**
7,738 + 2,022 ↓ ↓ + = _____	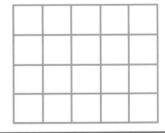
b. **Estimate:**	**Calculate exactly:**
9,152 − 4,728 ↓ ↓ − = _____	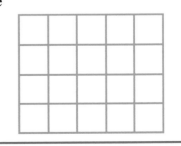

6. Fill in the tables.

a. Two thousand two				**b.** One thousand fifteen				**c.** Five thousand nine hundred six			
thou-sands	hund-reds	tens	ones	thou-sands	hund-reds	tens	ones	thou-sands	hund-reds	tens	ones

7. Write an addition or subtraction using an unknown, such as <u>?</u> or some symbol. Solve.

a. Hannah has saved $25. She wants to buy a bike for $69.
How much more does she still need to save?

b. Ashley bought a gift for her mom for $29, and now she has $16 left.
How much money did Ashley have before buying the gift?

8. Solve.

a. Carol bought six apples for $2.68,
a gallon of milk for $4.99, and a dozen
eggs for $2.95. Calculate her total bill.

b. Find the cost of three tickets for an airplane
flight if one ticket costs $267.

Also, estimate the answer using rounded
numbers.

My estimate: _____

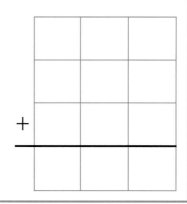

Geometry Review

1. **a.** Find the rhombi among these figures.

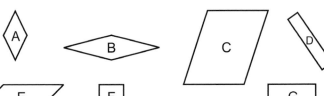

   b. Find quadrilaterals that are neither rectangles nor rhombi.

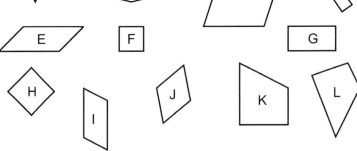

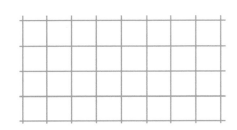

2. Draw a quadrilateral that is not a rectangle.

3. Fill in.

a. Write a multiplication for the area of this figure.	**b.** Draw a rectangle that has the area shown by the multiplication.
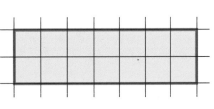	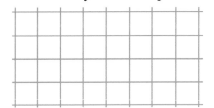
___ units × ___ units = ____ square units	4 × 5 = 20 square units

4. Find the perimeter and area of this rectangle. Use a centimeter ruler.

   Area:

   Perimeter:

5. Find the area and perimeter of these figures.

a. 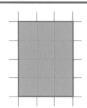 Area:

Perimeter:

b. 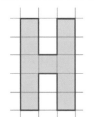 Area:

Perimeter:

6. Write a multiplication _and_ addition for the areas of these figures.

a.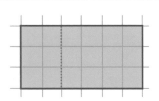

A = _____

b.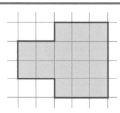

A = _____

7. Multiply using the shortcut.

| **a.** $7 \times 70 = $ _____ | **b.** $6 \times 80 = $ _____ | **c.** $40 \times 7 = $ _____ |

8. Find the total area of this rectangle, and the area of each part.

Area of each part:

Total area:

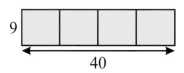

9. Draw and fill in.

a. Fill in the missing parts, and then draw a two-part rectangle to illustrate this number sentence.

$3 \times (5 + 1)$ = ___ × ___ + ___ × ___

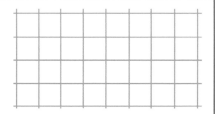

b. Fill in the missing parts, and then draw a two-part rectangle to illustrate this number sentence.

___ × (___ + ___) = $4 \times 2 + 4 \times 3$

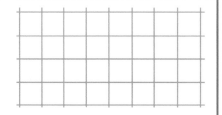

Geometry Test

1. Name any special quadrilaterals. If the quadrilateral does not have any special name, leave the line empty.

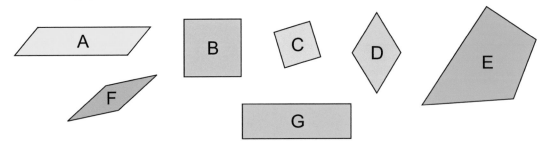

A _____

B _____

C _____

D _____

E _____

F _____

G _____

2. Find the area and perimeter of this figure.

Area = _____

Perimeter = _____

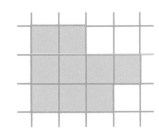

3. Solve. Write an addition with an unknown (?).

The perimeter of this rectangle is 42 cm. Its one side is 14 cm. How long is the other side?

_____

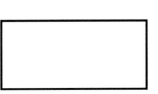

14 cm

?

Solution: _?_ = _____

48

4. Find the area and perimeter of these rectangles.

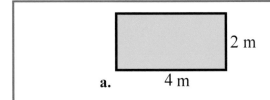

a.   4 m   2 m

Perimeter = _____

Area = _____

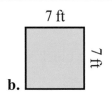

7 ft   7 ft

b.

Perimeter = _____

Area = _____

5. Write two multiplications to find the total area.

____ × ____ + ____ × ____ = _____

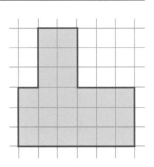

6. Margie's lawn is in the L-shape shown on the right. Calculate its area.

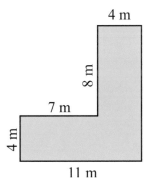

4 m

8 m

7 m

4 m

11 m

7. Jorge is planning to build a pen for his sheep. One possible pen would be a 60 ft by 80 ft rectangle, and the other possible pen would be a 40 ft by 120 ft rectangle.
   Which pen has a larger perimeter? How much larger?

8. Write a number sentence for the total area, thinking of one rectangle or two.

____ × (____ + ____) = ____ × ____ + ____ × ____

area of the
whole rectangle

area of the
first part

area of the
second part

Mixed Review 11

1. Write the time the clock shows.

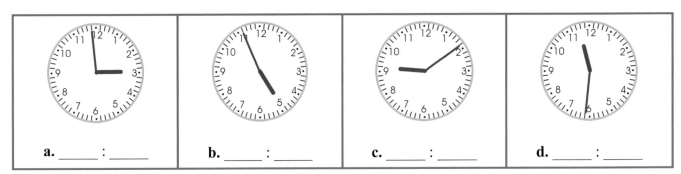

a. ____ : ____

b. ____ : ____

c. ____ : ____

d. ____ : ____

2. Subtract.

a.
$$\begin{array}{r} 4\ 5\ 6 \\ -\ 1\ 6\ 3 \\ \hline \end{array}$$

b.
$$\begin{array}{r} 7\ 2\ 1 \\ -\ 2\ 5\ 5 \\ \hline \end{array}$$

c.
$$\begin{array}{r} 4\ 8\ 0\ 2 \\ -\ 2\ 3\ 1\ 6 \\ \hline \end{array}$$

d.
$$\begin{array}{r} 3\ 7\ 0\ 0 \\ -\ 1\ 5\ 3\ 8 \\ \hline \end{array}$$

3. Solve. Below the addition, write a matching subtraction problem so that the numbers in the boxes are the same. Use mental math.

a. 99 + ☐ = 145

____ − ____ = ☐

b. 34 + ☐ = 76

____ − ____ = ☐

4. Compare. Write < , > , or = in the box.

a. 800 + 4000 ☐ 5000 + 400 + 80

b. 3000 + 60 + 5 ☐ 365

c. 20 + 8000 ☐ 4 + 8000 + 200

d. 400 + 9000 + 8 ☐ 80 + 900 + 8

e. 1 + 500 + 3000 ☐ 50 + 3000 + 900 + 9

f. 200 + 6000 + 40 + 7 ☐ 600 + 7000 + 2

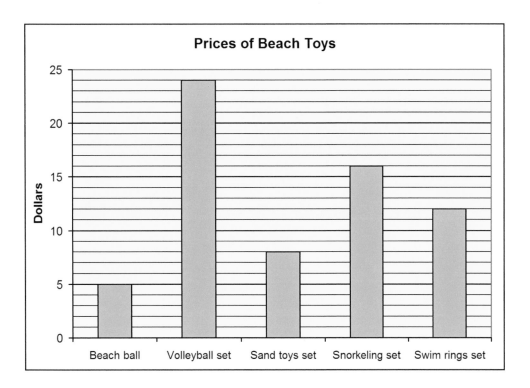

5. **a.** How much does the volleyball set cost?

   b. How much more does the snorkeling set
      cost than the swim rings set?

   c. How much do the sand toys and
      beach ball cost together?

   d. What is the total cost if you buy
      the two cheapest items?

6. Find the change.

a. A notebook costs $2.55. You give $3. Change: $_____	b. A book costs $5.88. You give $10. Change: $_____	c. A toy costs $6.70. You give $10. Change: $_____

7. Add parentheses to each equation to make it true.

a. $10 - 40 - 30 = 0$	b. $4 + 5 \times 2 - 1 = 17$	c. $5 \times 7 - 3 - 1 = 19$

Mixed Review 12

1. Write the following numbers in Roman numerals.

a. 16	b. 88	c. 149	d. 219

2. Multiply.

a.	b.	c.	d.
$5 \times 5 =$ _____	$2 \times 11 =$ _____	$2 \times 7 =$ _____	$5 \times 3 =$ _____
$12 \times 12 =$ _____	$8 \times 6 =$ _____	$4 \times 12 =$ _____	$1 \times 10 =$ _____
$7 \times 5 =$ _____	$3 \times 11 =$ _____	$6 \times 7 =$ _____	$8 \times 8 =$ _____

3. Diana bought two dolls for $3.35 each and two teddy bears for $6.90 each.

   a. Find the total cost.

   b. She paid with $25. How much was her change?

4. Round these numbers to the nearest hundred.

a.	b.	c.
$8,539 \approx$ _____	$9,687 \approx$ _____	$5,323 \approx$ _____
$3,551 \approx$ _____	$1,621 \approx$ _____	$2,399 \approx$ _____

5. Solve (find the number that the triangle stands for).

a. $1,500 + \triangle = 2,100$ $\triangle =$ _____	b. $5,200 - \triangle = 4,700$ $\triangle =$ _____	c. $\triangle - 2,300 = 2,300$ $\triangle =$ _____

6. Write the time the clock shows. Below, write the time using "past" and "till."

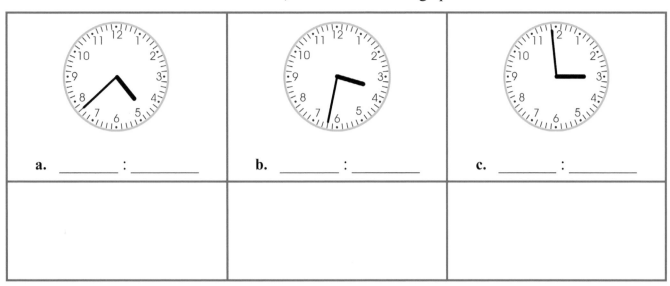

a. _____ : _____   b. _____ : _____   c. _____ : _____

7. Which math operations will make the number sentences true?

a. 14 ☐ (1 ☐ 12) = 26   c. 20 ☐ 4 ☐ 8 = 88

b. 90 ☐ 5 ☐ 4 ☐ 4 = 69   d. 10 ☐ (2 ☐ 4) ☐ 5 = 55

8. Gail got up at 5:29 and spent 3 minutes brushing her teeth,
   13 minutes showering and dressing, and 18 minutes eating
   breakfast. Then, she left for work.
   What time did she leave for work?

9. Find the missing money amounts.

a. $1.70 + _____ = $5	b. $2.29 + _____ = $3	c. $4.70 + _____ = $20
99 ¢ + _____ = $5	$3.47 + _____ = $10	$14.10 + _____ = $50

10. Children are watching videos. The problems below give you the starting time and the
    length of each video. Write the ending time.

a.	b.	c.
6:25 → _____ : _____	1:03 → _____ : _____	12:30 → _____ : _____
43 minutes	28 minutes	55 minutes

Measuring Review

1. Draw lines of these lengths:

   a. 4 1/4 in

   b. 5 cm 7 mm

2. Measure the sides of this triangle in centimeters and millimeters, and find its perimeter.

   AB: _____ cm _____ mm

   BC: _____ cm _____ mm

   CA: _____ cm _____ mm

   perimeter: _____ cm _____ mm

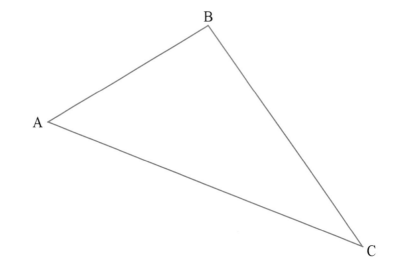

3. Measure the sides of this quadrilateral to the nearest quarter inch, and find its perimeter.

   AB: _____ in   BC: _____ in

   CD: _____ in   DA: _____ in

   perimeter: _____ in

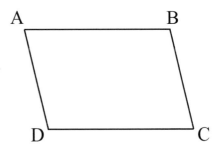

4. Write or say in order from the smallest to the biggest unit: cm km m mm

5. Write or say in order from the smallest to the biggest unit: ft in yd mi

6. Write or say in order from the smallest to the biggest unit: gal pt C qt

7. Name two different units you can use to measure the weight of people.

8. Fill in the blanks with suitable units of length. Sometimes several different units are possible

    a. A butterfly's wings were 6 _____ wide.   **b.** Sherry is 66 _____ tall.

    c. Jessica jogged 5 _____ yesterday.   **d.** The box was 60 _____ tall.

    e. The distance from the city   **f.** The room was 4 _____ wide.
       to the little town is 80 _____ .

                                            g. The eraser is 3 _____ long

9. Write the weight the scales are showing.

    a. _____ lb _____ oz   **b.** _____ lb _____ oz   **c.** _____ lb _____ oz

10. Have your teacher give you a small object. Use the scale to find out how much it weighs in either pounds and ounces, or in grams.

    It weighs _____.

11. Have your teacher give you a small container. Use a measuring cup to find out how much water it can hold in milliliters.

    It holds _____ ml.

12. Fill in the blanks with suitable units of weight and volume. Sometimes several different units are possible

    a. Mom bought 5 _____ of apples.   **b.** Mary drank 350 _____ of juice.

    c. Dr. Smith weighs about 70 _____ .   **d.** The banana weighed 3 _____ .

    e. The pan holds 2 _____ of water.   **f.** A cell phone weighs about 100 _____ .

Measuring Test

1. Draw lines of these lengths:

   a. 3 3/4 in

   b. 6 cm 5 mm

2. Measure the sides of this triangle in centimeters and millimeters.

3. Fill in each blank with a suitable unit. Sometimes several different units are possible

a. Mary's book weighed 350 _____.	**d.** The recipe called for 2 _____ of flour.
b. A juice box had 2 _____ of juice.	**e.** Mom bought 3 _____ of bananas.
c. The airplane was flying 10,000 _____ above the ground.	**f.** Andy and Matt bicycled 10 _____ to the beach.
g. Erika weighs 55 _____. **h.** The shampoo bottle can hold 450 _____ of shampoo. **i.** The large tank holds 200 _____ of water.	**j.** From Jerry's house to the neighbor's is 50 _____. **k.** A cell phone weighs 4 _____. **l.** A housefly measured 17 _____ long.

4. Write the units in order from the smallest to the biggest unit: ft in mi yd

5. Write the weight the scales are showing.

a. _____ lb _____ oz   **b.** _____ lb _____ oz

Mixed Review 13

1. Estimate the answers by rounding the numbers to the nearest ten or nearest hundred. Then find the exact answers.

a. A desk costs $154 and chairs cost $128. First, estimate the total cost. Then find the exact total.

My estimate: about $_____

b. Ed bought two computers for $1,298 each. First, estimate the total cost. Then find the exact total.

My estimate: about $_____

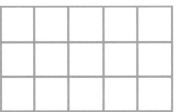

c. One TV costs $1,255 and another costs $787. Estimate the price difference between the two. Then find the exact difference.

My estimate: about $_____

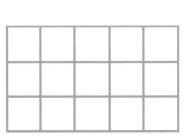

2. A rectangular pathway is 90 feet long and 6 feet wide. What is its area?

3. A large room is 20 feet by 9 feet. It is then divided into two equal parts. What is the area of one part?

4. Write the numbers immediately before and after the given number.

a. _____, 2,778, _____

b. _____, 6,060, _____

c. _____, 7,150, _____

d. _____, 7,000, _____

5. Multiply.

a. $5 \times 6 =$ _____	**b.** $6 \times 7 =$ _____	**c.** $9 \times 9 =$ _____
$3 \times 6 =$ _____	$4 \times 7 =$ _____	$8 \times 8 =$ _____
$8 \times 9 =$ _____	$5 \times 12 =$ _____	$6 \times 9 =$ _____
$7 \times 7 =$ _____	$8 \times 12 =$ _____	$6 \times 12 =$ _____

6. Break each multiplication into another where you multiply three numbers, one of them being 10. Multiply and fill in.

a. 7×30	**b.** 5×60
$=$ ____ $\times$ ____ $\times 10$	$=$ ____ $\times$ ____ $\times 10$
$=$ _____ $\times 10 =$ _____	$=$ _____ $\times 10 =$ _____

7. Multiply using the shortcut.

a. $8 \times 70 =$ _____	**b.** $3 \times 80 =$ _____	**c.** $50 \times 4 =$ _____
d. $30 \times 9 =$ _____	**e.** $20 \times 6 =$ _____	**f.** $4 \times 90 =$ _____

8. Find the total area of this rectangle, and also the area of each little part.

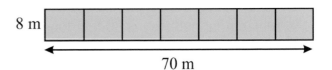

8 m   70 m

Total area = _____

Area of each part = _____

9. Subtract. Check by adding.

a. $\quad\begin{array}{r} 7\;2\;6\;2 \\ -\;2\;3\;1\;6 \\ \hline \end{array}$ + _____	**b.** $\quad\begin{array}{r} 6\;0\;0\;3 \\ -\;3\;2\;4\;2 \\ \hline \end{array}$ + _____

Mixed Review 14

1. Write an addition and a subtraction to match the bar model. Fill in the missing parts.

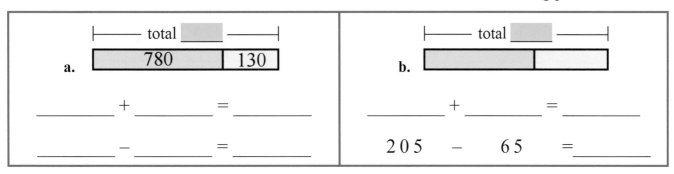

a.
total _____

| 780 | 130 |

_____ + _____ = _____

_____ – _____ = _____

b.
total _____

_____ + _____ = _____

2 0 5   –   6 5   = _____

2. Subtract. Check your answers.

a.
```
   7 9 0 4
 – 3 2 9 7      +  _____
_____
```

b.
```
   5 0 1 2
 – 3 2 7       +  _____
_____
```

3. Trisha needs three whole weeks to write a report. If she starts writing on November 3rd, when will she finish writing?

4. Compare. Write < , > , or = in the box.

a. 9,018 ☐ 9,180

b. 5,000 + 600 ☐ 500 + 6,000

c. 2,387 ☐ 2,378

d. 8,000 + 50 + 2 ☐ 200 + 5,000 + 800

5. Solve.

a. $10 \times 25 =$	b. $8 \times 90 =$	c. $70 \times 4 =$
d. $120 + 5 \times 7 =$	e. $12 \times 9 + 20 =$	f. $(11 - 3) \times 3 + 5 =$

6. Underline the greatest number in each box. Round each number to the nearest thousand.

a. 8,509 ≈ _____	**b.** 3,899 ≈ _____	**c.** 5,549 ≈ _____
5,479 ≈ _____	3,809 ≈ _____	5,459 ≈ _____
7,330 ≈ _____	3,890 ≈ _____	5,594 ≈ _____

7. In the grid on the right, draw a rectangle that is three units by four units. Then find its perimeter and area.

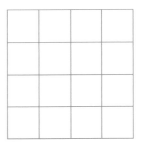

Perimeter: _____

Area: _____

8. Write a number sentence for the total area, thinking of one rectangle or two.

____ × (____ + ____) = _____ × _____ + _____ × _____

    area of the   area of the   area of the
whole rectangle   first part   second part

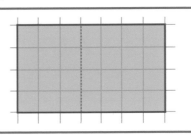

9. Now it's your turn to draw the rectangle. Fill in.

3 × (1 + 5) = _____ × _____ + _____ × _____

    area of the   area of the   area of the
whole rectangle   first part   second part

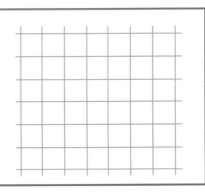

10. Write the names of these solids (three-dimensional figures).

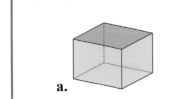

a. _____

b. _____

c. _____

d. _____

Division Review

1. Write a multiplication and a division fact to match the picture.

a. _____ × _____ = _____

   _____ ÷ _____ = _____

b. _____ × _____ = _____

   _____ ÷ _____ = _____

2. Divide.

a.	b.	c.	d.
36 ÷ 6 = _____	44 ÷ 11 = _____	56 ÷ 7 = _____	0 ÷ 9 = _____
3 ÷ 3 = _____	60 ÷ 6 = _____	72 ÷ 9 = _____	16 ÷ 16 = _____
36 ÷ 3 = _____	25 ÷ 5 = _____	99 ÷ 9 = _____	12 ÷ 1 = _____
4 ÷ 1 = _____	54 ÷ 9 = _____	100 ÷ 10 = _____	12 ÷ 2 = _____

3. Make fact families.

a.	b.	c.
_____ × 6 = 42	_____ × _____ = _____	_____ × _____ = _____
_____ × _____ = _____	_____ × _____ = _____	_____ × _____ = _____
_____ ÷ _____ = _____	_____ ÷ 8 = 1	_____ ÷ _____ = _____
_____ ÷ _____ = _____	_____ ÷ _____ = _____	49 ÷ _____ = 7

4. Find the missing numbers.

a. _____ × 5 = 45	b. _____ ÷ 5 = 4	c. _____ ÷ 3 = 3	d. 72 ÷ _____ = 8
e. 8 × _____ = 96	f. 56 ÷ 8 = _____	g. 54 ÷ _____ = 9	h. _____ ÷ 8 = 8

5. Multiply. Then for each multiplication, write two matching divisions **if possible**.
   Some divisions are not possible!

a. $6 \times 0 = \underline{\hspace{1cm}}$	**b.** $1 \times 9 = \underline{\hspace{1cm}}$	**c.** $0 \times 0 = \underline{\hspace{1cm}}$
$\underline{\hspace{1cm}} \div \underline{\hspace{1cm}} = \underline{\hspace{1cm}}$	$\underline{\hspace{1cm}} \div \underline{\hspace{1cm}} = \underline{\hspace{1cm}}$	$\underline{\hspace{1cm}} \div \underline{\hspace{1cm}} = \underline{\hspace{1cm}}$
$\underline{\hspace{1cm}} \div \underline{\hspace{1cm}} = \underline{\hspace{1cm}}$	$\underline{\hspace{1cm}} \div \underline{\hspace{1cm}} = \underline{\hspace{1cm}}$	$\underline{\hspace{1cm}} \div \underline{\hspace{1cm}} = \underline{\hspace{1cm}}$

6. Divide and find the remainder.

a. $11 \div 2 = \underline{\hspace{1cm}} R \underline{\hspace{1cm}}$	**b.** $41 \div 8 = \underline{\hspace{1cm}} R \underline{\hspace{1cm}}$	**c.** $16 \div 5 = \underline{\hspace{1cm}} R \underline{\hspace{1cm}}$
d. $56 \div 10 = \underline{\hspace{1cm}} R \underline{\hspace{1cm}}$	**e.** $26 \div 4 = \underline{\hspace{1cm}} R \underline{\hspace{1cm}}$	**f.** $22 \div 9 = \underline{\hspace{1cm}} R \underline{\hspace{1cm}}$

7. Solve the word problems. Write a division or a multiplication for each problem.

a. The teacher bought six boxes of crayons with eight in each box. How many crayons does she have?	**b.** The coach of a swimming club put 24 children into groups of six. How many groups did that make?
$\underline{\hspace{1cm}} \boxed{} \underline{\hspace{1cm}} = \underline{\hspace{1cm}}$	$\underline{\hspace{1cm}} \boxed{} \underline{\hspace{1cm}} = \underline{\hspace{1cm}}$
c. Rachel packaged cookies in bags to sell them. She had 48 cookies and she put 6 cookies in each bag. How many bags of cookies did she have?	**d.** Harry has put 94 stamps in his stamp book with ten stamps on each page. How many pages are full of stamps?
$\underline{\hspace{1cm}} \boxed{} \underline{\hspace{1cm}} = \underline{\hspace{1cm}}$	$\underline{\hspace{1cm}} \boxed{0} \underline{\hspace{1cm}} = \underline{\hspace{1cm}}$

Puzzle Corner

What numbers can go into the puzzles?

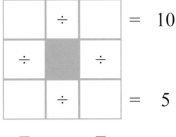

Division Test

1. For each multiplication, write two matching division facts.

a. $6 \times 7 = $ _____	**b.** $5 \times 11 = $ _____
_____ ÷ _____ = _____	_____ ÷ _____ = _____
_____ ÷ _____ = _____	_____ ÷ _____ = _____

2. Draw a picture to illustrate
   the division $20 ÷ 4 = 5$.

3. Divide.

a.	b.	c.	d.
$48 ÷ 6 = $ _____	$99 ÷ 11 = $ _____	$49 ÷ 7 = $ _____	$0 ÷ 3 = $ _____
$12 ÷ 3 = $ _____	$70 ÷ 7 = $ _____	$54 ÷ 9 = $ _____	$18 ÷ 18 = $ _____

4. Divide and find the remainder.

a. $60 ÷ 8 = $ ____ R ____	**b.** $73 ÷ 10 = $ ____ R ____	**c.** $36 ÷ 5 = $ ____ R ____

5. Solve the problems.

a. Fifty-four children in first grade are arranged into groups of 6 for a trip. How many groups will they make?	**b.** The teacher bought 4 packages of six markers and 4 packages of ten markers. How many markers did she buy?
c. Ashley has 85 stickers. She puts them in a notebook, nine stickers on each page. How many pages *full* of stickers will she get?	**d.** Ashley tore three pages full of stickers off her notebook. How many stickers are on those pages?

Mixed Review 15

1. Find the area and perimeter of this rectangle. Use a ruler to measure its sides.

   Area:

   Perimeter:

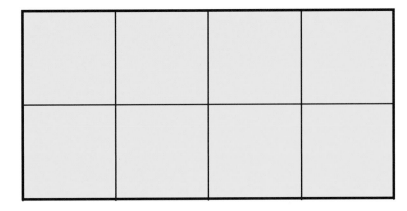

2. Write these numbers in order from smallest to greatest.

   2513   5096   5606   5060   2466   2506   2516

   _____ < _____ < _____ < _____ < _____ < _____ < _____

3. Fill in.

   + 20   + 100   + 300   + 40   + 60   + 600

   6880   _____ _____ _____ _____ _____   8000

4. Write using Roman numerals.

a. 8	b. 19	c. 40	d. 90
12	24	44	76

5. Round these numbers to the nearest hundred.

a.	b.	c.	d.
416 ≈ _____	529 ≈ _____	670 ≈ _____	254 ≈ _____
837 ≈ _____	960 ≈ _____	557 ≈ _____	147 ≈ _____

6. Solve the word problems.

a. One refrigerator costs $245 and another costs $68 less than that. Find the cost of the cheaper refrigerator. Also, estimate it using rounded numbers.

My estimate: about $_____

b. Mr. Sandman bought two of the cheaper refrigerators, and paid with $400. What was the total cost?

What was his change?

7. Draw lines using a ruler.

   a. 7 cm 8 mm

   b. 10 cm 5 mm

   c. 2 1/2 inches

   d. 4 3/4 inches

8. Fill in the blanks, using the units in, ft, or mi.

   a. Ann's living room is 20 _____ wide.

   b. The refrigerator is 28 _____ wide.

   c. It is about 2 _____ to the bookstore.

   d. The doctor is 6 _____ tall.

9. Fill in the blanks, using the units cm, km, mm, and m.

   a. The fly was 12 _____ long.

   b. The room measures about 3 _____ .

   c. Mark bicycled 12 _____ to go home.

   d. The teddy bear was 25 _____ tall.

Mixed Review 16

1. Solve (find the number that the symbol stands for).

a. $\triangle + 11 = 349$ $\triangle =$ _____	b. $530 - \triangle = 320$ $\triangle =$ _____	c. $\triangle - 1{,}600 = 500$ $\triangle =$ _____

2. Write an addition or a subtraction using an unknown, such as ? or some symbol. Solve.

a. The perimeter of a rectangle is 30 m. Its one side is 6 m.
   How long is the other side?

b. Mom had $200 when she went grocery shopping. She came back
   home with $78. How much did she spend in the store?

3. Write the numbers using Roman numerals.

a. 124	**b.** 40	**c.** 90	**d.** 222

4. Ronny is building a clubhouse. The floor will be seven feet by eight feet.

   a. What is the area of the floor?

   b. What is the perimeter of the floor?

5. Multiply.

a. $9 \times 5 =$ ___ $6 \times 5 =$ ___ $8 \times 5 =$ ___	b. $11 \times 12 =$ ___ $9 \times 12 =$ ___ $12 \times 12 =$ ___	c. $9 \times 9 =$ ___ $7 \times 9 =$ ___ $6 \times 9 =$ ___	d. $8 \times 7 =$ ___ $4 \times 7 =$ ___ $7 \times 7 =$ ___

6. Write a number sentence for the total area, thinking of one rectangle or two.

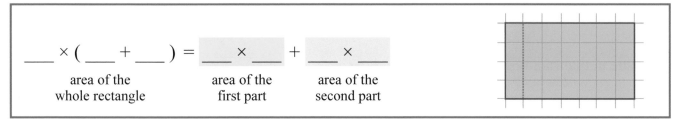

___ × (___ + ___) = ___ × ___ + ___ × ___

area of the whole rectangle   area of the first part   area of the second part

7. Write the cent amounts as dollar amounts, and vice versa.

a. _____ ¢ = $4.66   **b.** 3 ¢ = $_____   **c.** _____ ¢ = $2.05

8. Write the weight the scale shows using pounds and ounces.

      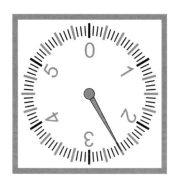

a. _____ lb _____ oz   **b.** _____ lb _____ oz   **c.** _____ lb _____ oz

9. Solve the problems.

a. Shelly has three favorite cookie recipes. The first recipe makes 5 dozen cookies, the second recipe makes 4 dozen, and the third recipe makes 2 dozen. How many cookies will Shelly have if she makes all three recipes?

b. Jim bought nine pairs of socks for $5 each, a shirt for $28, and pants for $47. What was the total cost?

c. Hal has two nickels and fifteen pennies. He traded them to Bill for one coin of equal value. What was the coin?

Fractions Review

1. Shade in the fractions. Then compare, and write >, <, or = between them.

a.	b.	c.
$\dfrac{2}{9}$ $\dfrac{2}{10}$	$\dfrac{5}{7}$ $\dfrac{5}{9}$	$\dfrac{1}{4}$ $\dfrac{1}{3}$

2. Divide the shapes into equal parts. Shade *one* part. Write the area of that part as a fraction of the whole area.

a. Divide the shape into seven equal parts.	**b.** Divide the shape into five equal parts.
	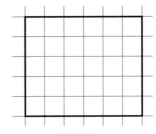
shaded area = ⬜/⬜ of the whole area	shaded area = ⬜/⬜ of the whole area

3. Write the fraction shown by the big dot on the number line.

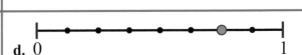

a. 0   1

b. 0   1

c. 0   1

d. 0   1

4. Mark the fraction on the number line (with a dot).

a. $\dfrac{5}{8}$	0 1	b. $\dfrac{2}{9}$	0 1

5. Write all the fractions under the tick marks.

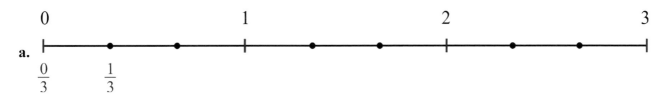

6. Write these fractions as mixed numbers or whole numbers. Use the number line.

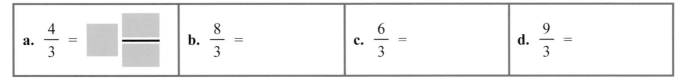

7. Write the whole numbers as fractions.

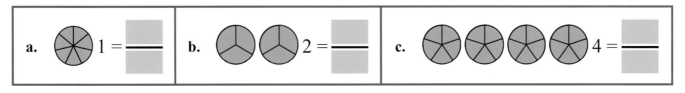

8. Draw pictures to illustrate these mixed numbers.

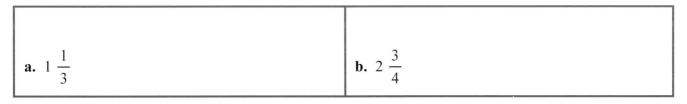

9. Use the fraction bars on the right to write two fractions that are equivalent to 1/3.

10. Write these fractions in order from smallest to largest. You can use the fraction bars to help you.

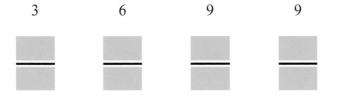

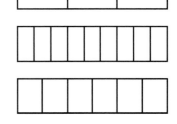

11. Compare the fractions. Write $<$, $>$ or $=$ between them.

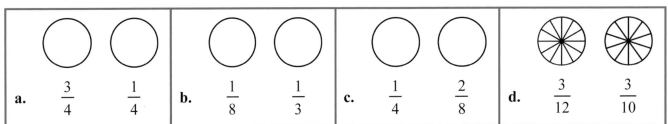

| **a.** $\dfrac{3}{4}$ $\dfrac{1}{4}$ | **b.** $\dfrac{1}{8}$ $\dfrac{1}{3}$ | **c.** $\dfrac{1}{4}$ $\dfrac{2}{8}$ | **d.** $\dfrac{3}{12}$ $\dfrac{3}{10}$ |

12. Compare the fractions. Write $<$, $>$ or $=$ between them.

a. $\dfrac{6}{8}$ □ $\dfrac{7}{8}$ **b.** $\dfrac{1}{5}$ □ $\dfrac{1}{10}$ **c.** $\dfrac{2}{9}$ □ $\dfrac{2}{5}$ **d.** $\dfrac{1}{2}$ □ $\dfrac{2}{4}$

13. Explain how to find which is the greater fraction: $\dfrac{8}{9}$ or $\dfrac{5}{9}$?

_____

14. Mark the equivalent fractions on the number lines.

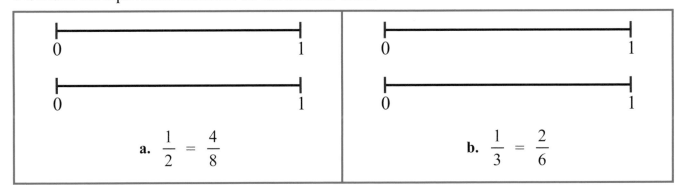

a. $\dfrac{1}{2} = \dfrac{4}{8}$ **b.** $\dfrac{1}{3} = \dfrac{2}{6}$

15. Write and shade the equivalent fractions.

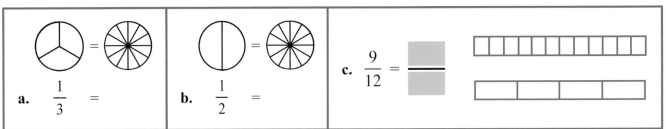

a. $\dfrac{1}{3} =$ **b.** $\dfrac{1}{2} =$ **c.** $\dfrac{9}{12} =$

16. Margaret drew these two pictures to show that $\dfrac{2}{8} = \dfrac{2}{4}$.

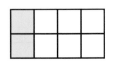

What do you think? Is she correct?

Fractions Test

1. Compare the fractions, and write >, <, or = .

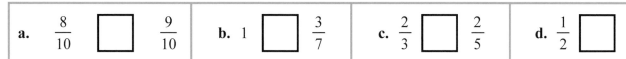

a. $\dfrac{8}{10}$ □ $\dfrac{9}{10}$	**b.** 1 □ $\dfrac{3}{7}$	**c.** $\dfrac{2}{3}$ □ $\dfrac{2}{5}$	**d.** $\dfrac{1}{2}$ □ $\dfrac{8}{9}$

2. Write these four fractions in order from the smallest to the largest: $\dfrac{1}{2}$ $\dfrac{1}{8}$ $\dfrac{1}{4}$ $\dfrac{1}{5}$

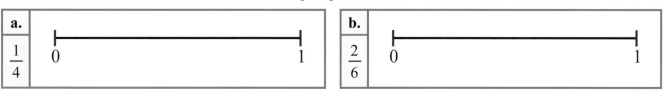

3. Divide the number line from 0 to 1 into equal parts. Then mark the fraction on it.

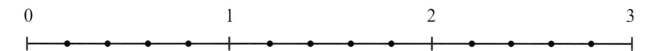

4. Mark the fractions and mixed numbers on the number line: $\dfrac{9}{5}$, $\dfrac{13}{5}$, $1\dfrac{1}{5}$, $2\dfrac{4}{5}$, $\dfrac{15}{5}$.

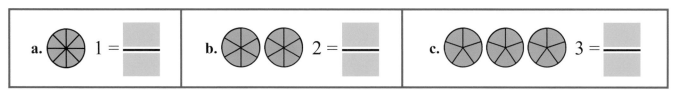

5. Write the whole numbers as fractions.

6. Write and shade the equivalent fractions.

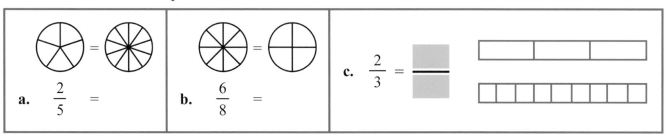

7. A loaf of bread is cut into 12 equal pieces.
   Another, similar loaf, is cut into 8 pieces.

   Mike ate 3 pieces of the first loaf.
   Janet ate 2 pieces of the second loaf.
   Who ate more bread?

8. This picture is trying to show that $\dfrac{3}{9} = \dfrac{3}{8}$.

   Explain why it is wrong.

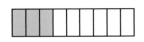

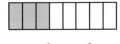

$$\frac{3}{9} = \frac{3}{8}$$

72

Mixed Review 17

1. Divide.

a.	b.	c.	d.
56 ÷ 7 = _____	48 ÷ 6 = _____	54 ÷ 9 = _____	48 ÷ 8 = _____
49 ÷ 7 = _____	72 ÷ 6 = _____	81 ÷ 9 = _____	72 ÷ 8 = _____
28 ÷ 7 = _____	54 ÷ 6 = _____	36 ÷ 9 = _____	32 ÷ 8 = _____

2. Write matching division and multiplication sentences.

a. _____ × _____ = _____	b. 3 × 0 = _____	c. _____ × _____ = _____
42 ÷ 7 = _____	_____ ÷ _____ = _____	_____ ÷ _____ = _____
_____ ÷ _____ = _____	_____ ÷ _____ = _____	72 ÷ 8 = _____

3. Divide and show the remainder.

a.	b.	c.
16 ÷ 5 = _____ R _____	21 ÷ 4 = _____ R _____	19 ÷ 6 = _____ R _____
12 ÷ 5 = _____ R _____	27 ÷ 4 = _____ R _____	31 ÷ 6 = _____ R _____

4. Kathy needs to read a 27-page booklet in three days. If she reads the same amount each day, how many pages will she read each day?

5. Six children are sharing 20 apples equally.
   How many apples will each child get?
   How many apples will be left over?

6. Write a number sentence for the shaded area and solve.

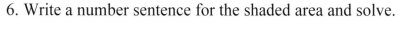

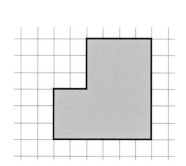

7. Round these numbers to the nearest ten, and estimate the perimeter of this park.

137 feet

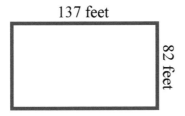

82 feet

Estimate: _____ ft

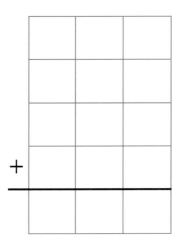

Also find the real perimeter by adding the original numbers in columns.

8. Write the fraction that the arrow points to on the number line.

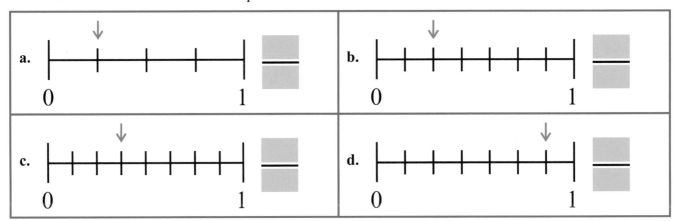

9. Explain how to find which is the greater fraction: $\frac{7}{10}$ or $\frac{7}{8}$?

_____

_____

10. One of the three numbers fits on the empty line so that the comparisons are true. Which number? Circle the number.

a. 5,637 5,673 5,607	b. 6,142 6,121 6,211
5,609 < _____ < 5,650	6,114 < _____ < 6,140
c. 6,996 9,966 9,696	d. 4,001 4,010 4,011
9,595 < _____ < 9,700	4,001 < _____ < 4,011

Mixed Review 18

1. Write the fact families.

a.	b.	c.
_____ × 2 = 14	_____ × _____ = _____	_____ × _____ = _____
_____ × _____ = _____	_____ × _____ = _____	_____ × _____ = _____
_____ ÷ 2 = _____	35 ÷ _____ = 7	42 ÷ 6 = _____
_____ ÷ _____ = _____	_____ ÷ _____ = _____	_____ ÷ _____ = _____

2. Solve.

a. Alice is bagging apples, four apples in each bag.
She has 12 bags. How many apples will she need?

b. Sofia had a piece of material that was 36 inches long. She cut
it into four equal pieces. How long is each piece?

c. How much money do you have if you have
five quarters, 3 dimes, and 22 pennies?

d. Write a word problem that will be solved with the division $24 \div 6 =$ _____

3. Find the missing numbers.

a. $40 \div$ _____ $= 8$	**b.** _____ $\div 5 = 7$	**c.** $4 \times$ _____ $= 48$	**d.** $4 \times 7 =$ _____
$72 \div$ _____ $= 8$	_____ $\div 6 = 7$	$8 \times$ _____ $= 48$	$8 \times 8 =$ _____

4. Solve.

a. $7 \div 2 =$ _____ , R _____	**b.** $13 \div 5 =$ _____ , R _____	**c.** $21 \div 2 =$ _____ , R _____
$9 \div 2 =$ _____ , R _____	$14 \div 5 =$ _____ , R _____	$47 \div 6 =$ _____ , R _____

5. **a.** What measuring units can you use to measure the weight of light items, such as an apple or a notebook?

   b. What measuring units can you use to measure the weight of heavy items, such as a refrigerator or a car?

   c. What measuring units can you use to measure the length of a room?

   d. What measuring units can you use to measure the height of a mountain?

6. The picture shows Amanda's garden. She is going to plant potatoes in the smaller part and different vegetables in the bigger part.

   a. Calculate the area that will be used for potatoes.

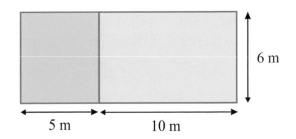

   b. Find the total area of her garden.

   c. Find the perimeter of the whole garden.

7. Mark the fractions on the number lines.

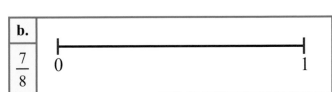

End-of-the-Year Test - Grade 3

This test is quite long, so I do not recommend having your child/student do it in one sitting. Break it into parts and administer them either on consecutive days, or perhaps on morning/evening/morning. This is to be used as a diagnostic test. You may even skip those areas that you already know for sure your student has mastered.

The test does not cover every single concept that is covered in the *Math Mammoth Grade 3 Complete Curriculum,* but all the major concepts and ideas are tested here. This test is evaluating the child's ability in the following content areas:

- multiplication tables and basic division facts
- mental addition and subtraction
- regrouping in addition and subtraction
- basic word problems
- multiplication and related concepts
- clock to the minute and elapsed time calculations
- basic money calculations (finding totals and change)
- place value and rounding with four-digit numbers
- quadrilaterals, perimeter, and area
- division and related concepts (remainder, word problems)
- measuring lines in inches and centimeters
- basic usage of measuring units
- the concept of a fraction and mixed number, equivalent fractions, and comparing fractions

Note 1: problems #2 and #3 are done <u>orally and timed</u>. Let the student see the problems. Read each problem aloud, and wait a maximum of 5-6 seconds for an answer. Mark the problem as right or wrong according to the student's (oral) answer. Mark it wrong if there is no answer. Then you can move on to the next problem.

You do not have to mention to the student that the problems are timed or that he/she will have 5-6 seconds per answer, because the idea here is not to create extra pressure by the fact it is timed, but simply to check if the student has the facts memorized (quick recall). You can say for example (vary as needed):

"I will ask you some multiplication and division questions. Try to answer me as quickly as possible. In each question, I will only wait a little while for you to answer, and if you do not say anything, I will move on to the next problem. So just try your best to answer the questions as quickly as you can."

In order to continue with the Math Mammoth Grade 4 Complete Curriculum, I recommend that the child gain a minimum score of 80% on this test, and that the teacher or parent review with him any content areas that are found weak. Children scoring between 70 and 80% may also continue with grade 4, depending on the types of errors (careless errors or not remembering something, vs. lack of understanding). The most important content areas to master are the multiplication tables and the word problems, because of the level of logical reasoning needed in them. Use your judgment.

My suggestion for grading is below. The total is 207 points. A score of 166 points is 80%.

Grading on question 1 (the multiplication tables grid): There are 169 empty squares to fill in the table, and the completed table is worth 17 points. Count how many of the answers the student gets right, divide that by 10, and round to the nearest whole point. For example: a student gets 24 right. 24/ 10 = 2.4, which rounded becomes 2 points. Or, a student gets 85 right. 85/10 = 8.5, which rounds to 9 points.

Question	Max. points	Student score
Multiplication Tables and Basic Division Facts		
1	17 points	
2	16 points	
3	16 points	
	subtotal	/ 49
Addition and Subtraction, Including Word Problems		
4	6 points	
5	6 points	
6	4 points	
7	4 points	
8	4 points	
9	3 points	
10	3 points	
11	4 points	
	subtotal	/ 34
Multiplication and Related Concepts		
12	1 point	
13	1 point	
14	3 points	
15	3 points	
16	1 point	
17	2 points	
18	1 point	
	subtotal	/ 12
Time		
19	8 points	
20	3 points	
	subtotal	/ 11

Question	Max. points	Student score
Graphs		
21a	1 point	
21b	1 point	
21c	1 point	
21d	2 points	
	subtotal	/ 5
Money		
22a	1 point	
22b	2 points	
22c	2 points	
23	2 points	
24	3 points	
	subtotal	/ 10
Place Value and Rounding		
25	2 points	
26	5 points	
27	4 points	
28	2 points	
29	8 points	
	subtotal	/ 21
Geometry		
30	5 points	
31	2 points	
32	4 points	
33	2 points	
34	2 points	
35	3 points	
	subtotal	/ 18

Question	Max. points	Student score
Measuring		
36	2 points	
37	2 points	
38	2 points	
39	6 points	
	subtotal	/ 12
Division and Related Concepts		
40	2 points	
41	6 points	
42	3 points	
43	2 points	
44	2 points	
	subtotal	/ 15
Fractions		
45	6 points	
46	3 points	
47	2 points	
48	3 points	
49	4 points	
50	2 points	
	subtotal	/ 20
	TOTAL	**/ 207**

End-of-the-Year Test Grade 3

Multiplication Tables and Basic Division Facts

1. Fill in the complete multiplication table.
   You have 12 minutes to fill it in completely.

×	0	1	2	3	4	5	6	7	8	9	10	11	12
0													
1													
2													
3													
4													
5													
6													
7													
8													
9													
10													
11													
12													

In problems 2 and 3, your teacher will read you multiplication and division questions. Try to answer them as quickly as possible. In each question, he/she will only wait a little while for you to answer, and if you do not say anything, your teacher will move on to the next problem. So just try your best to answer the questions as quickly as you can.

2. Multiply.

a.	b.	c.	d.
$2 \times 7 = $ _____	$7 \times 4 = $ _____	$3 \times 3 = $ _____	$7 \times 8 = $ _____
$8 \times 3 = $ _____	$5 \times 8 = $ _____	$4 \times 4 = $ _____	$6 \times 5 = $ _____
$5 \times 5 = $ _____	$3 \times 9 = $ _____	$7 \times 7 = $ _____	$8 \times 6 = $ _____
$9 \times 4 = $ _____	$5 \times 7 = $ _____	$4 \times 8 = $ _____	$6 \times 9 = $ _____

3. Divide.

a.	b.	c.	d.
$21 \div 3 = $ _____	$32 \div 4 = $ _____	$45 \div 5 = $ _____	$50 \div 5 = $ _____
$35 \div 7 = $ _____	$40 \div 8 = $ _____	$28 \div 4 = $ _____	$72 \div 9 = $ _____
$48 \div 6 = $ _____	$66 \div 6 = $ _____	$36 \div 9 = $ _____	$18 \div 6 = $ _____
$49 \div 7 = $ _____	$56 \div 8 = $ _____	$63 \div 7 = $ _____	$27 \div 9 = $ _____

Addition and Subtraction, including Word Problems

4. Add in your head and write the answers.

a. $240 + 70 =$ _____	**b.** $540 + 80 =$ _____	**c.** $59 + 89 =$ _____
$99 + 50 =$ _____	$335 + 9 =$ _____	$46 + 34 =$ _____

5. Subtract in your head and write the answers.

a. $100 - 67 =$ _____	**b.** $651 - 8 =$ _____	**c.** $52 - 37 =$ _____
$73 - 68 =$ _____	$54 - 9 =$ _____	$400 - 22 =$ _____

6. Subtract. Then check your answer using the grid.

a.

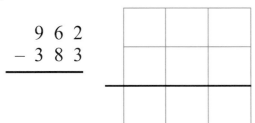

$$\begin{array}{r} 9\ 6\ 2 \\ -\ 3\ 8\ 3 \\ \hline \end{array}$$

b.

$$\begin{array}{r} 7\ 0\ 0\ 2 \\ -\ 4\ 5\ 2\ 6 \\ \hline \end{array}$$

7. Solve.

a. $82 + 5{,}539 + 1{,}254 + 278$

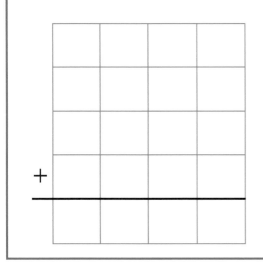

b. $535 + (430 - 173)$

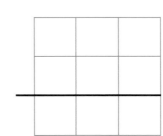

8. Solve what number goes in place of the triangle.

a. $414 +$ $= 708$

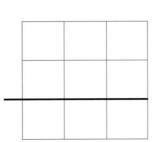 is _____

b. $- 339 = 485$

is _____

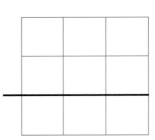

Solve.

9. Jason bought a $185 camera and a $32 camera bag.
   What was his change from $300?

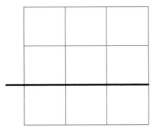 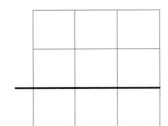

10. A family is driving 300 miles from their hometown to Grandma's place.
    10 miles before the half-way point they stopped to have lunch.
    How many miles do they still have to go?

11. A store received 100 boxes, which each had 8 light bulbs.

   a. How many light bulbs did the store receive?

   b. After selling eight boxes, how many bulbs are left?

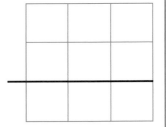

Multiplication and Related Concepts

12. Draw a picture to illustrate
    the multiplication $3 \times 4 = 12$.

13. Solve: $5 \times 25 =$ _____

14. Solve.

a. $24 + 8 \times 3$	b. $2 + (5 + 4) \times 2$	c. $66 - 5 \times 5$

15. Write a multiplication sentence (NOT just the answer) to solve how many legs these
    animals have in total.

    a. Seven horses: _____

    b. Five ducks: _____

    c. Eight horses and six ducks: _____

16. Each table in a restaurant seats four people. How many
    tables do you need to reserve for a party of 31 people?

17. A cafeteria menu had spaghetti with meatballs
    for $8 and bean soup for $6. How much would
    it cost to buy three plates of spaghetti with meatballs
    and three bowls of bean soup?

18. Anna is bagging hair clips she made. She will put four hair clips in each bag.
    She has 28 hair clips to bag. How many bags will she need?

Time

19. Write the time the clock shows, and the time 10 minutes later.

	a. _____ : _____	b. _____ : _____	c. _____ : _____	d. _____ : _____
10 min. later	_____ : _____	_____ : _____	_____ : _____	_____ : _____

20. **a.** The TV show starts at 6:25 PM and ends at 7:10 PM.
    How long is it?

   b. Mr. Jackson's plane takes off at 9:30 AM. If the flight
   lasts for 6 hours 20 minutes, when will the plane land?

   c. The baseball game was going to be on May 21,
   but it was postponed (made later) by one week.
   What was the new date for the game?

Graphs

21. The graph shows some
    people's working hours
    on Uncle Ted's apple farm.

   a. How many hours did Erica work?

   b. How many hours did Kathy work?

   c. How many more hours did Jason
   work than Jack?

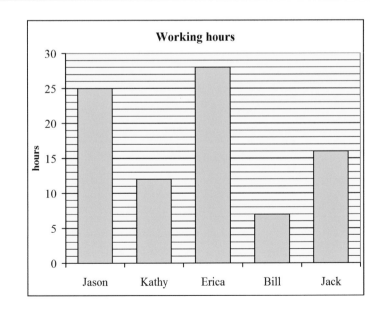

   d. How many hours did the three boys work in total?

Money

22. Find the total cost of buying the items listed. Line up the numbers carefully when you add.

$6.60	$8.95	$1.25	$16.59

a. a calculator and a bag	**b.** two pens and a book	**c.** three pens and a calculator

23. Find the change.

a. A book costs $7.10. You give $10. Change: $_____	**b.** A basket costs $4.45. You give $5. Change: $_____

24. A pencil case costs $2.35. If Mark buys four of them with his $10, what will his change be?

Place Value and Rounding

25. Fill in the missing numbers.

a. $2,000 + 60 +$ _____ $= 2,760$	**b.** $700 + 20 +$ _____ $+ 9 = 2,729$

26. Compare and write $<$, $>$, or $=$.

a. $6,034$ ☐ $3,064$	**b.** $5,156$ ☐ $5,516$	**c.** $9,079$ ☐ $9,097$
d. $6,000 + 3 + 40$ ☐ $400 + 60 + 3,000$	**e.** $900 + 7,000$ ☐ $90 + 7,000 + 2$	

27. Add and subtract.

a. $5,400 + 300 =$ _____	**b.** $2,900 - 1,700 =$ _____
$7,800 + 800 =$ _____	$8,100 - 300 =$ _____

28. Round the numbers to the nearest <u>TEN</u>.

a. $743 \approx$ _____	**b.** $987 \approx$ _____	**c.** $251 \approx$ _____	**d.** $665 \approx$ _____

29. Estimate these calculations by rounding the numbers to the nearest <u>hundred</u>. Also, calculate the exact answer.

a. Round the numbers, then add: $3,782 \quad + \quad 2,255$ $\downarrow \qquad\qquad \downarrow$ $+ \qquad\qquad =$ _____	**Calculate exactly:** $\begin{array}{r} 3\ 7\ 8\ 2 \\ +\ 2\ 2\ 5\ 5 \\ \hline \end{array}$
b. Round the numbers, then subtract: $8,149 \quad - \quad 888$ $\downarrow \qquad\qquad \downarrow$ $- \qquad\qquad =$ _____	**Calculate exactly:** $\begin{array}{r} 8\ 1\ 4\ 9 \\ -\ \ \ 8\ 8\ 8 \\ \hline \end{array}$

30. Name any special quadrilaterals.

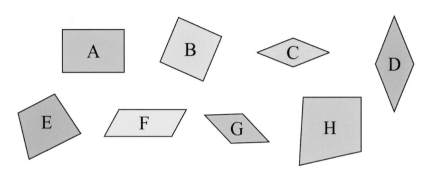

_____

_____

_____

_____

_____

_____

31. Find the perimeter and area of this shape.

Perimeter: _____

Area : _____

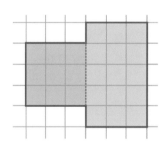

32. The picture shows a two-part lawn.

a. Find the areas of part 1 and part 2.

_____ and _____

b. Find the perimeter of the whole lawn.

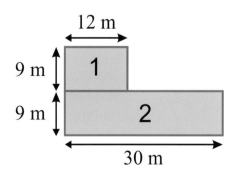

33. The perimeter of a rectangle measures 26 in. Find the other side length, if one side measures 4 in.

34. Draw in the grid below:

a. a rectangle with an area of 15 square units

b. a rectangle with a perimeter of 10 units.

35. Write a number sentence for the total area, thinking of one rectangle or two.

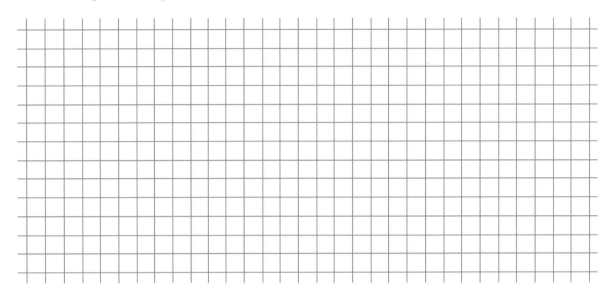

___ × (___ + ___) = ___ × ___ + ___ × ___ = _____

area of the whole rectangle   area of the first part   area of the second part

Measuring

36. Draw lines:

a. 6 1/4 inch long

b. 7 cm 5 mm long

37. Write in order from smallest to biggest unit: cm km m mm

38. Name two different units that you can use to measure a small
    amount of water in a drinking glass.

39. Fill in the blanks with units of measure. Sometimes several different units are possible.

a. The mountain is 20,000 _____ high.

b. The pencil is 14 _____ long.

c. Jeremy bought 5 _____ of potatoes.

d. The large glass holds 3 _____ of liquid.

e. The teacher weighs 68 _____ .

f. The room was 20 _____ wide.

Division and Related Concepts

40. Write two multiplications and two divisions for the same picture.

_____ × _____ = _____   _____ ÷ _____ = _____

_____ × _____ = _____   _____ ÷ _____ = _____

41. Divide, but CROSS OUT all the problems that are impossible!

a. $17 \div 1 =$ _____	**b.** $17 \div 17 =$ _____	**c.** $1 \div 1 =$ _____
$17 \div 0 =$ _____	$0 \div 0 =$ _____	$0 \div 1 =$ _____

42. Divide.

a. $17 \div 2 =$ _____ R _____   **b.** $24 \div 5 =$ _____ R _____   **c.** $47 \div 7 =$ _____ R _____

43. A team leader divided a group of 24 children into teams.
    Can he divide the children equally into teams of 5?
    Teams of 6?   Teams of 7?

44. Annie, Rob, and Ted decided to buy a gift that cost $16 and flowers that cost $14 for Mom.
    The children shared the total cost equally. How much did each child pay?

Fractions

45. Write the fraction or mixed number.

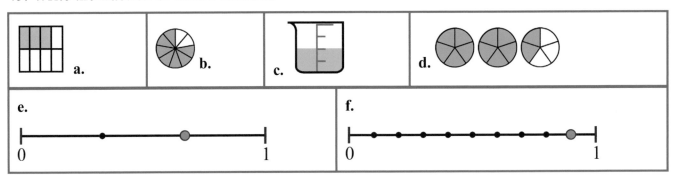

46. Write the whole numbers as fractions.

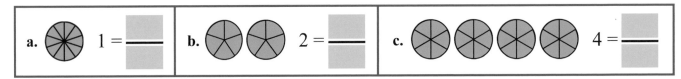

47. Mark the equivalent fractions $\frac{3}{6}$ and $\frac{1}{2}$ on the number lines.

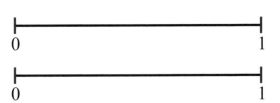

48. Shade parts for the first fraction. Shade the same *amount* in the second picture, forming an equivalent fraction. Write the second fraction.

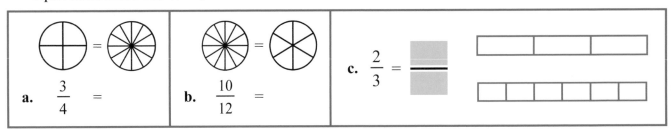

49. Compare the fractions, and write > , < , or = in the box.

a. $\frac{2}{7}$ □ $\frac{2}{3}$ b. $\frac{5}{11}$ □ $\frac{7}{11}$ c. $\frac{1}{2}$ □ $\frac{9}{10}$ d. $\frac{1}{7}$ □ $\frac{1}{8}$

50. Mary ate 1/2 of a strawberry pie, and David ate 7/12 of a blueberry pie. Look at the pictures. Who ate more pie?

Mary's pie: David's pie:

Math Mammoth Grade 3 Review Workbook Answers

Addition and Subtraction Review, p. 6

1. a. 308, 304   b. 230, 465   c. 994, 198

2. a. VI = 6	b. LVI = 56	c. LXV = 65	d. XLVIII = 48
e. 8 = VIII	f. 14 = XIV	g. 23 = XXIII	h. 67 = LXVII

3. a. 139   b. 294   c. 378   d. 377   e. 166

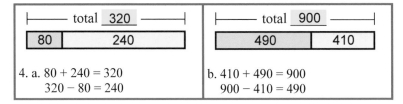

4. a. 80 + 240 = 320
   320 − 80 = 240

  b. 410 + 490 = 900
   900 − 410 = 490

5.

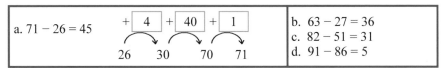

a. 71 − 26 = 45

b. 63 − 27 = 36
c. 82 − 51 = 31
d. 91 − 86 = 5

6. a. 31   b. 41   c. 510   d. 390

7. a. 12 ≈ 10   b. 677 ≈ 680   c. 46 ≈ 50

8. a. 27   b. 785

9. a. 800 − 270 − 270 = 260 yellow beads
  b. 100 + 100 + 100 + (100 − 14) = 386 CDs

Addition and Subtraction Test, p 8

1. a. 270; 203   b. 93;129   c. 47; 871

2. a. 5   b. 287   c. 8

3. a. 4   b. 66   c. 78   d. 144   e. 29   f. 98

4. <u>He has $105 left.</u> His purchases were a total of $145. And, $250 − $145 = $105.

5. a. <u>247</u>. To check it, add 247 + 157 = 404.
  b. <u>326</u>. To check it, add 326 + 397 = 723

6. a. 710   b. 600   c. 820   d. 460

7. a. 27   b. 43   c. 310   d. 320

8. 159 days

9. Jason has 4 × 80 − 28 = 320 − 28 = <u>292 trading cards.</u>

10.

```
|————— total 438 —————|
| 171 |      267       |
```

    171 + 267 = 438 (or 267 + 171 = 438)

    438 − 171 = 267 (or 438 − 267 = 171)

11. 295

1. a.    b.

2. a. 7 + 7 + 7 = 21
   b. 20 + 20 + 20 + 20 = 80

3. a. 5 × 4 = 20
   b. 9 × 3 = 27

4.

a. 2 × 2 = 4
    1 × 4 = 4
	b. 2 × 10 = 20
	    3 × 3 = 9
		c. 12 × 0 = 0
		    12 × 1 = 12

d. 0 × 5 = 0
    2 × 7 = 14
	e. 2 × 40 = 80
	    3 × 30 = 90
		f. 2 × 400 = 800
		    1 × 500 = 500

5. a. 20 balls
   b. 5 × 4 = 20 legs
   c. 7 − 2 = 5 × 2 = 10 cans of cat food.
   d. 7 − 1 = 6 × 3 + 1 = 19 books total

6. a. 10   b. 17   c. 23   d. 12

7.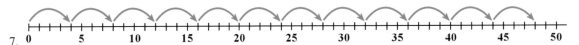

1 × 4 = 4	4 × 4 = 16	7 × 4 = 28	10 × 4 = 40
2 × 4 = 8	5 × 4 = 20	8 × 4 = 32	11 × 4 = 44
3 × 4 = 12	6 × 4 = 24	9 × 4 = 36	12 × 4 = 48

Concept of Multiplication Test, p. 12

1. a. 6, 5, 0
   b. 10, 30, 12
   c. 40, 120, 400
   d. 9, 0, 11

2. Answers will vary. Check the student's answers. For example:

    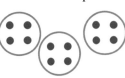

   a.   b.

3. a. There are 3 × 12 = <u>36 apples</u> in three baskets.
   b. 4 × $2 + 2 × $8 = $24
   c. You can make <u>five groups</u>. 5 × 4 = 20

4. a. 20   b. 22   c. 0   d. 11

Mixed Review 1, p. 13

1. a. 3, 7   b. 15, 23
  c. 9, 61   d. 24, 175

2.

a. $93 + 6 = 99$ $893 + 6 = 899$	b. $47 + 29 = 76$ $607 + 9 = 616$	c. $15 + 18 = 33$ $624 + 8 = 632$

3.

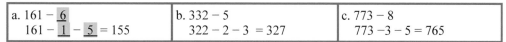

a. $161 - \underline{6}$ $161 - \underline{1} - \underline{5} = 155$	b. $332 - 5$ $322 - 2 - 3 = 327$	c. $773 - 8$ $773 - 3 - 5 = 765$

4.

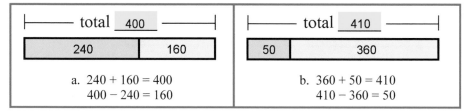

total 400 240 160 a. $240 + 160 = 400$ $400 - 240 = 160$	total 410 50 360 b. $360 + 50 = 410$ $410 - 360 = 50$

5.

a. $19 - (6 + 2) + 5 = 16$ $19 - 6 + 2 + 5 = 20$	b. $(800 - 60) - (50 - 40) = 730$ $800 - 60 - 50 - 40 = 650$

6. a. 259, 835
  b. 176, 602

7. a. Danny ran 735 yards.
  b. They still have 174 km to go.
  c. The two jars have 580 beans.

8.

a. a toy, $28, and a set of books, $129 toy about $30 set of books about $130 together about $160	b. a ladder, $62, and wheelbarrow, $137 ladder about $60 wheelbarrow about $140 together about $200

Mixed Review 2, p. 15

1. a. 574   b. 810   c. 983

2. a. XXV   b. XIX   c. LVII   d. CXLIII

3.

a. $35 - 14 - 7 + 3 = 17$ b. $35 - (14 - 7) + 3 = 31$ c. $35 - (14 - 8 + 3) = 26$	d. $(250 - 20) + (80 - 30) = 280$ e. $250 - (20 + 80 - 30) = 180$ f. $250 - 20 + (80 - 30) = 280$

4. Jill needs three more cups for her tea party.

5. a. 9, 8, 0
  b. 10, 18, 32
  c. 90, 80, 800
  d. 0, 0, 22

6. a. $20 + 20 = 40$
  b. $50 + 50 + 50 = 150$

Mixed Review 2, cont.

7. a. 12 + 12 + 12 = 36 *or* 3 × 12 = 36. Tim has <u>36 feet</u> of string.
   b. She needs <u>six</u> bags. 16 + 8 = 24; 6 × 4 = 24.

8.

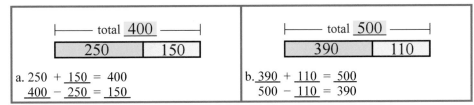

a. 250 + <u>150</u> = 400
   <u>400</u> − <u>250</u> = 150

b. <u>390</u> + <u>110</u> = <u>500</u>
   500 − <u>110</u> = 390

9. a. 589   b. 316   c. 258   d. 143

10. 48 + 48 + 48 = 144. Yes, you can buy three bicycles for $150 and have $6 left over.

11.

a. Rent, $256. Groceries, $387. Rent: about $260 Groceries: about $390 Total: about $650	b. Ticket for adult, $58. Ticket for child, $38. Ticket for adult: about $60 Ticket for child: about $40 Total cost: about $100

Multiplication Tables Review, p. 17

1.

×	0	1	2	3	4	5	6	7	8	9	10	11	12
0	0	0	0	0	0	0	0	0	0	0	0	0	0
1	0	1	2	3	4	5	6	7	8	9	10	11	12
2	0	2	4	6	8	10	12	14	16	18	20	22	24
3	0	3	6	9	12	15	18	21	24	27	30	33	36
4	0	4	8	12	16	20	24	28	32	36	40	44	48
5	0	5	10	15	20	25	30	35	40	45	50	55	60
6	0	6	12	18	24	30	36	42	48	54	60	66	72
7	0	7	14	21	28	35	42	49	56	63	70	77	84
8	0	8	16	24	32	40	48	56	64	72	80	88	96
9	0	9	18	27	36	45	54	63	72	81	90	99	108
10	0	10	20	30	40	50	60	70	80	90	100	110	120
11	0	11	22	33	44	55	66	77	88	99	110	121	132
12	0	12	24	36	48	60	72	84	96	108	120	132	144

2. a. 9 × 8 $<$ 10 × 8   b. 9 × 5 $>$ 11 × 4   c. 9 × 2 $=$ 3 × 6

   d. 9 × 8 $>$ 9 × 4   e. 4 × 4 $=$ 2 × 8   f. 10 × 11 $>$ 10 × 7

   g. 10 × 8 $>$ 10 × 5   h. 9 × 2 $<$ 4 × 5   i. 9 × 8 $>$ 9 × 6

Multiplication Tables Review, cont.

3.

$1 \times 3 = 3$	$7 \times 3 = 21$	$1 \times 6 = 6$	$7 \times 6 = 42$
$2 \times 3 = 6$	$8 \times 3 = 24$	$2 \times 6 = 12$	$8 \times 6 = 48$
$3 \times 3 = 9$	$9 \times 3 = 27$	$3 \times 6 = 18$	$9 \times 6 = 54$
$4 \times 3 = 12$	$10 \times 3 = 30$	$4 \times 6 = 24$	$10 \times 6 = 60$
$5 \times 3 = 15$	$11 \times 3 = 33$	$5 \times 6 = 30$	$11 \times 6 = 66$
$6 \times 3 = 18$	$12 \times 3 = 36$	$6 \times 6 = 36$	$12 \times 6 = 72$

Every other answer from the table of three is in the table of six.

4. a. $11 \times 7 = 77$ The girls have a total of 77 schoolbooks.
  b. $4 \times 5 = 20$ There will be five groups.
  c. $4 \times 3 + 7 = 19$ The total cost was $19.
  d. $12 \times 2 = 24$ He bought 12 packages of seed.
  e. $5 \times 4 + 3 \times 4 + 20 \times 2 = 72$ They have a total of 72 feet.

5. a. 3, 8, 5   b. 3, 11, 2   c. 7, 8, 9   d. 5, 9, 7
  e. 4, 7, 9   f. 12, 7, 9   g. 6, 4, 9   h. 5, 7, 9

Mystery numbers: a. 44.   b. 24   c. 29   d. 24   e. 44   f. 12

Multiplication Tables Test, p. 20

1.

×	0	1	2	3	4	5	6	7	8	9	10	11	12
0	0	0	0	0	0	0	0	0	0	0	0	0	0
1	0	1	2	3	4	5	6	7	8	9	10	11	12
2	0	2	4	6	8	10	12	14	16	18	20	22	24
3	0	3	6	9	12	15	18	21	24	27	30	33	36
4	0	4	8	12	16	20	24	28	32	36	40	44	48
5	0	5	10	15	20	25	30	35	40	45	50	55	60
6	0	6	12	18	24	30	36	42	48	54	60	66	72
7	0	7	14	21	28	35	42	49	56	63	70	77	84
8	0	8	16	24	32	40	48	56	64	72	80	88	96
9	0	9	18	27	36	45	54	63	72	81	90	99	108
10	0	10	20	30	40	50	60	70	80	90	100	110	120
11	0	11	22	33	44	55	66	77	88	99	110	121	132
12	0	12	24	36	48	60	72	84	96	108	120	132	144

2. a. $3 \times 8 = 24$ (or $8 \times 3 = 24$)
  b. $9 \times 7 = 63$ (or $7 \times 9 = 63$)

3. a. $5 \times \$9 + 5 \times \$5 = \$70$.
  b. You need nine tables. $9 \times 6 = 54$.
  c. They have $7 \times 4 + 4 \times 4 + 12 \times 2 = 68$ feet in total.
  d. You can buy 8 shirts. $8 \times \$6 = \$48$.

4. a. 4, 9, 6   b. 11, 6, 2   c. 7, 4, 11   d. 9, 4, 12
  e. 6, 11, 2   f. 8, 2, 4   g. 6, 9, 4   h. 12, 4, 7

Mixed Review 3, p. 22

1. a. 27   b. 687   c. 5

2. a. 660   b. 600   c. 820   d. 60

3. a. 746   b. 721

4. 16 + 36 = 52   or 52 − 16 = 36. There are 36 white candles.

5.

| a. 12 is XII | b. 34 is XXXIV | c. 55 is LV | d. 80 is LXXX |

6. a. △ = 27   b. △ = 700   c. △ = 430

7.

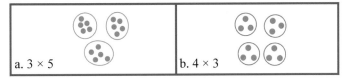

8. a. 5 × 10 = 50   Five children have 50 toes all totaled.
   b. 3 × 5 = 15   He has three rows of cars.
   c. 5 × 4 = 20   You need five tables to seat 20 people.

9. a. The Sports Club is the most popular.
   b. There are 15 more students.
   c. There are 68 students in the three clubs.

Mixed Review 4, p. 24

1. Jimmy rode 44 miles each day.

2. They traveled a total of 469 miles.

3.

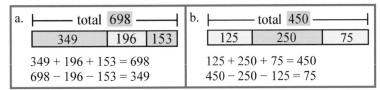

4. a. 4 × 8 + 10 × 2 = 52. The total cost was $52.
   b. 12 ÷ 4 = 3 + 6 = 9. She spent $9.

5. a. 9   b. 121   c. 67   d. 14

6. a. 2 × 5 = 10, 5 × 2 = 10
   b. 2 × 7 = 14, 7 × 2 = 14

7. a. They saw 90 giraffes.   b. They saw 45 more giraffes.

Telling Time Review, p. 26

1. a. 11:51; 12:01   b. 8:43; 8:53   c. 4:57; 5:07   d. 1:14; 1:24

2. a. 15 min.   b. 35 min.   c. 38 min.   d. 34 min.

3. a. 6 hours   b. 14 hours   c. 18 min.   d. 40 min.

4. It ends at 2:35.

5. The trip was 1 hour 10 minutes long.

Telling Time Test, p. 27

1. a. 1:47;1:57   b. 10:09; 10:19   c. 5:34;5:44   d. 9:49; 9:59

2. a. 36 minutes   b. 39 minutes   c. 2 hours   d. 33 minutes.   e. 44 minutes   f. 33 minutes   g. 2 hours 20 minutes.

3. a. It ends at 8:30.   b. The trip was 3 hours 35 minutes.   c. 2 hours 15 minutes

Mixed Review 5, p. 28

1. a. 25   b. 43   c. 29   d. 389   e. 561   f. 803

2. a. 14   b. 66   c. 49   d. 140

3. $89 - 17 = 72 + 72 = 144$ or $2 \times 89 - 2 \times 17 = 144$; They cost $144.

4. a. 27, 28, 0   b. 24, 24, 24   c. 56, 63, 32   d. 36, 60, 21

5. a. $7 \times 2 = 14$; He read 14 books.
   b. $3 \times 7 + 5 = 26$; He put 26 pencils in the pencil cases.

6. If there is anything in parentheses, do it first. Do the multiplications before additions or subtractions.
   Then, do the additions and subtractions from left to right. The first step is highlighted.

   a. $2 + \boxed{5 \times 2} = 12$ b. $5 \times \boxed{(1 + 1)} = 10$ c. $\boxed{(4 - 2)} \times 7 = 14$

7. a. $22 + \underline{119} = 141$; Davy weighs 119 pounds.
   b. $275 - 48 = \underline{227}$; The cheaper washer costs $227.

8. $430 + 430 + 280 + 280 = 1{,}420$ meters approximately
   or $400 + 400 + 300 + 300 = 1{,}400$ meters approximately.

Mixed Review 6, p. 30

1.
a. $650 + 120 = 770$	b. $633 + 9 = 642$
$770 - 120 = 650$	$642 - 9 = 633$

2. a. 26   b. 56   c. 67

3. a. $13 - 9 = 4$; 16
   b. $250 - 50 = 200$; 140

4. a. <u>Yes.</u> $30 - 1 - 2 = 27$; so there are 27 cupcakes left. For the afternoon tea they need $2 \times (1 + 12) = 26$ cupcakes.
   b. <u>Nine</u> marbles were lost. $9 \times 12 = 108$; $108 - 99 = 9$.

5.
a. 6:08	b. 6:48	c. 3:29	d. 5:33
e. 4:36	f. 5:39	g. 11:58	h. 12:43

6.
a. $8 \times 10 - 2 + 5 = 83$	b. $6 + 7 \times (4 - 2) = 20$
c. $3 \times 4 - 2 \times 3 = 6$	d. $2 \times (4 + 4) \times 2 = 32$

7.
a. $564 - 5 = 559$	b. $888 + 12 = 900$
$564 - 10 = 554$	$886 + 14 = 900$
$564 - 15 = 549$	$884 + 16 = 900$
$564 - 20 = 544$	$882 + 18 = 900$
$564 - 25 = 539$	$880 + 20 = 900$
$564 - 30 = 534$	$878 + 22 = 900$

8. a. $3 \times 7 + 2 \times 7 = 35$. She spent a total of <u>35 days</u> at the beach and at the farm in total.
   b. $15 \times 5 \times 2 = 150$. He spends <u>150 minutes</u> walking to and from school.

Money Review, p. 32

1. a. $10.40   b. $7.56

2. a. $0.75   b. $1.73   c. $1.45

3. a. Maria still needs to save $19.95.
   b. Arnold's total bill is $6.70.
   c. His change is $3.30.

4. a. My total bill is $3.55.
   b. My change is $1.45.

Money Test, p. 33

1. a. $8.90   b. $2.06

2. a. $1.40   b. $1.21   c. $1.16

3. a. The total cost is $4.30.
   b. The change is $0.70 or 70 cents.

4. a. Marsha still needs to save $16.85.
   b. The total cost is $8.51.
   c. His change is $11.49.

Mixed Review 7, p. 34

1. a.

   b.
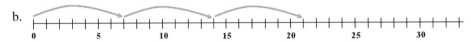

2. a. $50 - (20 - 7) = 37$
   b. $(8 - 5) \times 2 - 1 = 5$
   c. $(15 + 5) \times (2 - 1) = 20$ OR $15 + 5 \times (2 - 1) = 20$

3. a. 72, 49, 54   b. 48, 35, 28
   c. 36, 81, 72   d. 84, 64, 18

4. a. $5 \times 8 = 40$; You can buy eight pairs of socks.
   b. $7 \times 4 = 28$; There will be four layers of dominoes.

5.
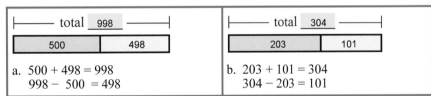

a. $500 + 498 = 998$
   $998 - 500 = 498$

b. $203 + 101 = 304$
   $304 - 203 = 101$

6. a. 3, 7, 9   b. 9, 6, 5
   c. 8, 6, 4   d. 6, 9, 7

7. a. The tickets would cost $142.
   b. There are 52 pages left to read.
   c. There are 85 blue ribbons.

Mixed Review 8, p. 36

1. a. 62   b. 61   c. 64

2. a. 32   b. 28   c. 56

3. a. $350 - 18$ $<$ $350 - 15$ b. $180 - 15$ $=$ $190 - 25$

   c. $264 + 7$ $<$ $267 + 8$ d. $62 - 27$ $>$ $61 - 27$

4. The potatoes cost $0.96 and your change is $4.04.

5. She spent 54 days in three states.

6. a. 577   b. 485

7. a. She sold about $100 + 60 + 25 + 30 = \underline{215}$ parrot magnets.
   b. She sold about $40 + 85 + 55 + 25 = \underline{205}$ other kinds of magnets.

8. a. 35   b. 30   c. 88

9. a. 7 hours 52 min   b. 3 hours 33 min   c. 2 hours 38 min

10. a. $0.30   b. $0.91   c. $0.80

Place Value with Thousands Review, p. 38

1. a. 7,240   b. 6,005   c. 2,029

2. a. 7,503; 3,090
   b. 1,037; 6,400

3. a. >   b. >   c. <   d. <

4. a. 1,900; 7,200
   b. 3,300; 3,700
   c. 800; 900
   d. 4,900; 8,300

5. a. △ = 700   b. △ = 9,800   c. △ = 1,500

6. a. 900   b. 5,300   c. 6,000   d. 2,700

7. a. Estimate: $2,500 + 1,800 = 4,300$, Exact: 4,343
   b. Estimate: $6,600 - 700 = 5,900$, Exact: 5,845

8. a. Estimate: $1,600 + $300 + 1,000 = $2,900$.   Exact: $2,863.
   b. Estimate: $5,000 - $300 - $1,300 = $3,400$. Exact: $5,000 - $278 - $1,250 = $3,472$.

Place Value with Thousands Test, p. 40

1. a. 2,689   b. 4,070   c. 5,609   d. 3,902

2. a. >   b. >   c. <   d. <

3. a. 700; 8,200
   b. 8,100; 8,100

4. a. 500   b. 1,400   c. 2,900

5. a. Estimate: $2,900 + 4,500 = 7,400$. Exact calculation: 7,396
   b. Estimate: $7,000 - 3,000 = 4,000$; Exact calculation: 4,029

6. a. Estimation: $2,000 - ($1,600 + $300) = 100. The exact answer: $86.

   b. Estimation: $2,600 - $700 = $1,900$. The exact answer: $1916.

Mixed Review 9, p. 42

1. a. 49, 249   b. 63, 663   c. 75, 275

2. a. $6 \times 2 + 4 \times 4 = 28$ legs
   b. $9 \times 4 + 3 = 39$ windows

3. a. 570, 340   b. 900, 260
   c. 600, 430   d. 60, 1,000

4. a. 3, 8, 14   b. 19, 24, 60
   c. 85, 53, 40   d. 43, 125, 271

5. a. XV, XIX   b. XXI, XLIII   c. LVI, LXV   d. XC, XCIX

6. a. 4, 9, 7
   b. 7, 5, 8
   c. 8, 6, 4
   d. 4, 8, 5

7. a. 5 hours 20 min|
   b. 4 hours 30 min
   c. 11 hours
   d. 5 hours

8. a. $4.40   b. $15.27   c. $27.64

Mixed Review 10, p. 44

1. a. 4, 8   b. 0, 9   c. 6, 9   d. 9, 9

2. a. XV   b. XXXII   c. XLVII   d. LVI

3. a. 22 till 7   b. 4 till 4   c. 12 past 2   d. 17 till 8

4. It ended at 8:18.

5. a. Estimate: $7,700 + 2,000 = 9,700$ Exact: 9,760
   b. Estimate: $9,200 - 4,700 = 4,500$ Exact: 4,424

6.

a. Two thousand two				b. One thousand fifteen				c. Five thousand nine hundred six			
thou-sands	hund-reds	tens	ones	thou-sands	hund-reds	tens	ones	thou-sands	hund-reds	tens	ones
2	0	0	2	1	0	1	5	5	9	0	6

7. a. $25 + ? = $69. OR $69 - $25 = ? The unknown ? = $44. She needs to save $44 more.

   b. ? - $29 = $16.   The unknown ? = $45. She had $45 before buying the gift.

8. a. $2.68 + $4.99 + $2.95 = 10.62
   b. Estimate: $3 \times 270 = 810$ Exact: $801

Geometry Review, p. 46

1. a. A, B, F, H, J   b. C, E, I, K, L

2. Answers vary. Check the student's answers.

3.

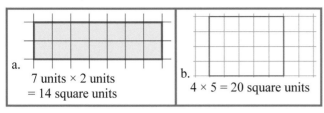

    a. 7 units × 2 units = 14 square units

    b. 4 × 5 = 20 square units

4. a. Area 35 cm^2
   b. perimeter 24 cm

5. a. Area 12 square units; perimeter 14 units
   b. Area 11 square units; perimeter 24 units

6. a. A = 3 × 2 + 3 × 4 = 18 square units
   b. A = 2 × 2 + 3 × 4 = 16 square units

7. a. 490   b. 480   c. 280

8. Area of each part: 9 × 10 = 90 square units. Total area 9 × 40 = 360 square units.

9.

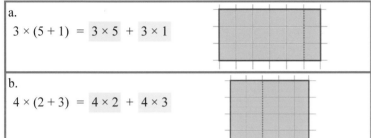

  a.

    3 × (5 + 1) = 3 × 5 + 3 × 1

  b.

    4 × (2 + 3) = 4 × 2 + 4 × 3

Geometry Test, p. 48

1. A -
  B a square
  C a square
  D a rhombus
  E -
  F a rhombus
  G a rectangle

2. Area = 9 square units. Perimeter = 14 units.

3. 14 cm + ? = 21 cm   or   14 cm + ? + 14 cm + ? = 42 cm   Solution: ? = 7 cm

4. In this problem the student needs to also give the correct *unit*, not just the correct number.
   a. Perimeter = 12 m. Area = 8m^2 b. Perimeter = 28 ft   c. Area = 49ft^2 or 49 square feet

5. 2 × 3 + 6 × 3 = 24 square units *or* 2 × 3 + 3 × 6 = 24 square units
   or 3 × 2 + 6 × 3 = 24 square units *or* 3 × 2 + 3 × 6 = 24 square units

6. Divide the shape into two rectangles (which can be done in two different ways).
   Area = 4 m × 11 m + 4 m × 8 m = 76 m^2 OR   7 m × 4 m + 12 m × 4 m = 76 m^2.
   The student needs to include square meters with his/her answer (m^2), not just the correct number.

7. The latter pen has a larger perimeter. Its perimeter is 320 ft, whereas the first pen's is 280 ft. The difference is 40 ft.
   It is required that the student include feet with his/her answer ft), not just the correct number.

8.   3 × (4 + 2) = 3 × 4 + 3 × 2
   area of the   area of the   area of the
   whole rectangle first part   second part

Mixed Review 11, p. 50

1. a. 2:59   b. 4:56   c. 9:09   d. 11:31

2. a. 293   b. 466   c. 2,486   d. 2,162

3.

a. 99 + ☐ = 145	b. 34 + ☐ = 76
145 − 99 = 46	76 − 34 = 42

4. a. <   b. >   c. <   d. >   e. <   f. <

5. a. The volleyball set costs $24.
   b. The snorkeling set costs $4 more than the swim ring set.
   c. They cost $13.
   d. The cheapest would be $13.

6. a. The change is $0.45.   b. The change is $4.12.   c. The change is $3.30.

7.

a. 10 − (40 − 30) = 0	b. (4 + 5) × 2 − 1 = 17	c. 5 × (7 − 3) − 1 = 19

Mixed Review 12, p. 52

1. a. XVI   b. LXXXVIII   c. CXLIX   d. CCXIX

2.

a.	b.	c.	d.
5 × 5 = 25	2 × 11 = 22	2 × 7 = 14	5 × 3 = 15
12 × 12 = 144	8 × 6 = 48	4 × 12 = 48	1 × 10 = 10
7 × 5 = 35	3 × 11 = 33	6 × 7 = 42	8 × 8 = 64

3. a. The total cost was $20.50.
   b. Her change was $4.50.

4.

a.	b.	c.
8,539 ≈ 8,500	9,687 ≈ 9,700	5,323 ≈ 5,300
3,551 ≈ 3,600	1,621 ≈ 1,600	2,399 ≈ 2,400

5. a. △ = 600   b. △ = 500   c. △ = 4,600

6.

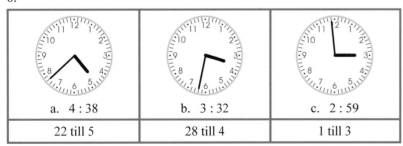

a. 4 : 38	b. 3 : 32	c. 2 : 59
22 till 5	28 till 4	1 till 3

7. a. 14 + (1 × 12) = 26   c. 20 × 4 + 8 = 88
   b. 90 − 5 − 4 × 4 = 69   d. 10 × (2 + 4) − 5 = 55

8. She left for work at 6:03.

9. a. $3.30, $4.01   b. $0.71, $6.53   c. $15.30, $35.90

10. a. 7:08   b. 1:31   c. 1:25

Measuring Review, p. 54

1. a. _____

   b. _____

2. AB: _5_ cm _1_ mm
   BC: _7_ cm _2_ mm
   CA: _9_ cm _2_ mm
   perimeter: _21_ cm _5_ mm

3. AB: _1 ½_ in   BC: _1_ in
   CD: _1 ½_ in   DA: _1_ in
   perimeter: _5_ in

4. mm, cm, m, km

5. in, ft, yd, mi

6. C, pt, qt, gal

7. pounds or kilograms

8. a. A butterfly's wings were 6 _cm_ wide.   b. Sherry is 66 _in_ tall.
   c. Jessica jogged 5 _km or mi_ yesterday.   d. The box was 60 _cm_ tall.
   e. The distance from the city   f. The room was 4 _m_ wide.
      to the little town is 80 _km or mi_ .   g. The eraser is 3 _cm_ long

9. a. 2 lb 12 oz   b. 2 lb 4 oz   c. 5 lb 12 oz

10. Answers will vary.

11. Answers will vary.

12. a. Mom bought 5 _kg or lb_ of apples.   b. Mary drank 350 _ml_ of juice.
   c. Dr. Smith weighs about 70 _kg_ .   d. The banana weighed 3 _oz_ .
   e. The pan holds 2 _qt or L_ of water.   f. A cell phone weighs about 100 _g_ .

Measuring Test, p. 56

1. a. _____

   b. _____

2.
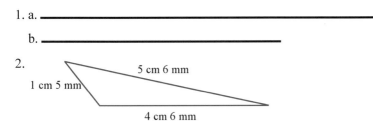

3.

a. Mary's book weighed 350 _g_ .
 b. A juice box had 2 _L / qt_ of juice.
 c. The airplane was flying 10,000 _ft / m_
    above the ground.
	d. The recipe called for 2 _C_ of flour.
	 e. Mom bought 3 _kg / lb_ of bananas.
	 f. Andy and Matt bicycled 10 _km / mi_
	    to the beach.

g. Erika weighs 55 _kg / lb_ .
 h. The shampoo bottle can hold
    450 _ml_ of shampoo.
 i. The large tank holds 200 _gal / L_ of water.
	j. From Jerry's house to the neighbor's
	    is 50 _ft / m_ .
	 k. A cell phone weighs 4 _oz_ .
	 l. A housefly measured 17 _mm_ long.

4. in ft yd mi

5. a. 1 lb 11 oz   b. 3 lb 9 oz

Mixed Review 13, p. 57

1. a. Estimate: $150 + $130 = $280. Exact: $154 + $128 = $282.
   b. Estimate: $1,300 + $1,300 = $2,600. Exact: $1,298 + $1,298 = $2,596.
   c. Estimate: $1,300 − $800 = $500. Exact: $1,255 − $787 = $468.

2. Area = $90 \times 6 = 540 \text{ ft}^2$

3. $20 \times 9 = 180$. Half of 180 is 90 so the area of one part is 90 ft².

4. a. 2,777, 2,778, 2,779   b. 6,059, 6,060, 6,061
   c. 7,149, 7,150, 7,151   d. 6,999, 7,000, 7,001

5.

a. $5 \times 6 = 30$	b. $6 \times 7 = 42$	c. $9 \times 9 = 81$
$3 \times 6 = 18$	$4 \times 7 = 28$	$8 \times 8 = 64$
$8 \times 9 = 72$	$5 \times 12 = 60$	$6 \times 9 = 54$
$7 \times 7 = 49$	$8 \times 12 = 96$	$6 \times 12 = 72$

6.

a. 7×30	b. 5×60
$= 7 \times 3 \times 10$	$= 5 \times 6 \times 10$
	$= 30 \times 10 = 300$

7.

a. $8 \times 70 = 560$	b. $3 \times 80 = 240$	c. $50 \times 4 = 200$
d. $30 \times 9 = 270$	e. $20 \times 6 = 120$	f. $4 \times 90 = 360$

8. The total area is 560 m². The area of each part is 80 m².

9. a. 4,946 Check: 4,945 + 2,316 = 7,262.
   b. 2,761 Check: 2,761 + 3,242 = 6,003.

Mixed Review 14, p. 59

1.

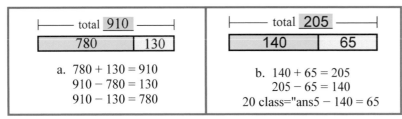

    a. 780 + 130 = 910
       910 − 780 = 130
       910 − 130 = 780

    b. 140 + 65 = 205
       205 − 65 = 140
    20 class="ans5 − 140 = 65

2. a. 4,607   b. 4,685

3. Since she started on November 3rd, she will finish on November 23rd. Notice that from November 3rd to November 23rd is 21 days, or three weeks. You will include both November 3rd and November 23rd in this count of 21 days.

4. a. <   b. <   c. >   d. >

5. a. 250   b. 720   c. 280   d. 155   e. 128   f. 29

6.

a.	b.	c.
8,509 ≈ 9,000	3,899 ≈ 4,000	5,549 ≈ 6,000
5,479 ≈ 5,000	3,809 ≈ 4,000	5,459 ≈ 5,000
7,330 ≈ 7,000	3,890 ≈ 4,000	5,594 ≈ 6,000

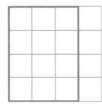

7. Perimeter 14 units.
   Area 12 square units.

8.

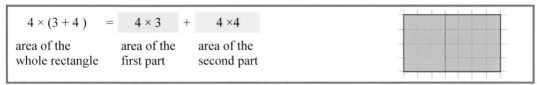

$4 \times (3 + 4)$	$=$	4×3	$+$	4×4
area of the whole rectangle		area of the first part		area of the second part

9.

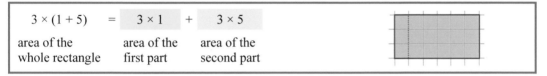

$3 \times (1 + 5)$	$=$	3×1	$+$	3×5
area of the whole rectangle		area of the first part		area of the second part

10. a. box   b. cube   c. cylinder   d. cone

Division Review, p. 61

1. $2 \times 6 = 12$ $12 \div 6 = 2$ b. $3 \times 5 = 15$ $15 \div 5 = 3$

2.

a.	b.	c.	d.
$36 \div 6 = 6$	$44 \div 11 = 4$	$56 \div 7 = 8$	$0 \div 9 = 0$
$3 \div 3 = 1$	$60 \div 6 = 10$	$72 \div 9 = 8$	$16 \div 16 = 1$
$36 \div 3 = 12$	$25 \div 5 = 5$	$99 \div 9 = 11$	$12 \div 1 = 12$
$4 \div 1 = 4$	$54 \div 9 = 6$	$100 \div 10 = 10$	$12 \div 2 = 6$

3.

a.	b.	c.
$7 \times 6 = 42$	$8 \times 1 = 8$	$7 \times 7 = 49$
$6 \times 7 = 42$	$1 \times 8 = 8$	$7 \times 7 = 49$
$42 \div 6 = 7$	$8 \div 1 = 8$	$49 \div 7 = 7$
$42 \div 7 = 6$	$8 \div 8 = 1$	$49 \div 7 = 7$

4. a. 9   b. 20   c. 9   d. 9   e. 12   f. 7   g. 6   h. 64

5.

a. $6 \times 0 = 0$	b. $1 \times 9 = 9$	c. $0 \times 0 = 0$
$0 \div 6 = 0$	$9 \div 1 = 9$	$\cancel{0 \div 0}$

6.

a. $11 \div 2 = 5$ R1	b. $41 \div 8 = 5$ R1	c. $16 \div 5 = 3$ R1
d. $56 \div 10 = 5$ R6	e. $26 \div 4 = 6$ R2	f. $22 \div 9 = 2$ R4

7. a. $6 \times 8 = 48$ She has 48 crayons.
   b. $24 \div 6 = 4$ There were four groups of six children.
   c. $48 \div 6 = 8$ She had eight bags of cookies.
   d. $94 \div 10 = 9$ R 4. (Or, $9 \times 10 + 4 = 94$), so 9 pages are full of stamps. The tenth page has 4 stamps on it.

Division Review, cont.

Puzzle corner:

80	÷	8	= 10
÷		÷	
10	÷	2	= 5
=		=	
8		4	

54	÷	9	= 6
÷		÷	
6	÷	3	= 2
=		=	
9		3	

Division Test, p. 63

1.

a. 6 × 7 = 42	b. 5 × 11 = 55
42 ÷ 7 = 6	55 ÷ 11 = 5
42 ÷ 6 = 7	55 ÷ 5 = 11

2.  or

3. a. 8,4   b. 9, 10   c. 7, 6   d. 0, 1

4. a. 7 R4   b. 7 R3   c. 7 R1

5. a. Nine groups. 9 × 6 = 54.
   b. 4 × 6 + 4 × 10 = 24 + 40 = 64 markers.
   c. Nine pages. 9 × 9 = 81, and 10 × 9 = 90.
   d. 27 stickers. 3 × 9 = 27

Mixed Review 15, p. 64

1. Area 8 in². Perimeter 12 in.

2. 2466 < 2506 < 2513 < 2516 < 5060 < 5096 < 5606

3.

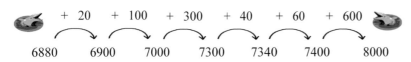

	+ 20	+ 100	+ 300	+ 40	+ 60	+ 600	
6880	6900	7000	7300	7340	7400	8000	

4. a. VIII, XII   b. XIX , XXIV   c. XL, XLIV   d. XC, LXXVI

5.

a.	b.	c.	d.
416 ≈ 400	529 ≈ 500	670 ≈ 700	254 ≈ 300
837 ≈ 800	960 ≈ 1000	557 ≈ 600	147 ≈ 100

6. a. Estimate: $250 − $70 = $180   Exact: $245 − $68 = $177
   b. The total cost was $354. $400 − $177 − $177 = $46 change.

Mixed Review 15, cont.

7. Check the student's answers.

   a. ————————————————————

   b. ————————————————————————

   c. ————————————————

   d. ——————————————————————————————

8. a. Ann's living room is 20 ft wide.   b. The refrigerator is 28 in wide.
   c. It is about 2 mi to the bookstore.   d. The doctor is 6 ft tall.

9. a. The fly was 12 mm long.   b. The room measures about 3 m.
   c. Mark bicycled 12 km to go home.   d. The teddy bear was 25 cm tall.

Mixed Review 16, p. 66

1. a. 338. Solve by subtracting $349 - 11 = \underline{338}$.
   b. 210. Solve by subtracting $530 - 320 = \underline{210}$.
   c. 2,100. Solve by adding $1,600 + 500 = 2,100$.

2. a. $30 \div 2 - 6 = ?$ or $6 + ? \times 2 = 30$. The other side is 9 meters long.
   b. $\$78 + ? = \$200$. She spent $122.

3. a. CXXIV   b. XL   c. XC   d. CCXXII

4. a. The area will be 56 ft^2. The perimeter will be 30 feet.

5. a. $9 \times 5 = 45$ b. $11 \times 12 = 132$ c. $9 \times 9 = 81$ d. $8 \times 7 = 56$
       $6 \times 5 = 30$ $9 \times 12 = 108$ $7 \times 9 = 63$ $4 \times 7 = 28$
       $8 \times 5 = 40$ $12 \times 12 = 144$ $6 \times 9 = 54$ $7 \times 7 = 49$

6.

$4 \times (1 + 6)$	=	4×1	+	4×6
area of the whole rectangle		area of the first part		area of the second part

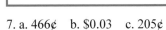

7. a. 466¢   b. $0.03   c. 205¢

8. a. 0 lb 8 oz   b. 0 lb 12 oz   c. 2 lb 8 oz

9. a. She will have 11 dozen cookies, which is a total of **132 cookies**.
   b. The total cost was $120.
   c. He traded his coins for one quarter.

1.

a. $\dfrac{2}{9} > \dfrac{2}{10}$ b. $\dfrac{5}{7} > \dfrac{5}{9}$ c. $\dfrac{1}{4} < \dfrac{1}{3}$

2.

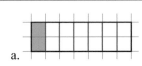

a. shaded area = $\dfrac{1}{7}$ of the whole area

b. shaded area = $\dfrac{1}{5}$ of the whole area

3. a. 1/5   b. 2/10   c. 8/9   d. 6/8

4. a.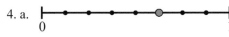

b.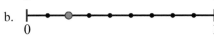

5.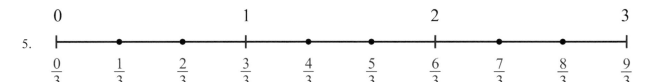

6. a. 1 1/3   b. 2 2/3   c. 2   d. 3

7. a. 1 = 7/7   b. 2 = 6/3   c. 4 = 20/5

8. a.    b.

9.

$\dfrac{1}{3} = \dfrac{3}{9} = \dfrac{2}{6}$

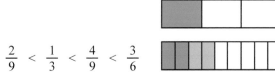

10.

$\dfrac{2}{9} < \dfrac{1}{3} < \dfrac{4}{9} < \dfrac{3}{6}$

Fractions Review, cont.

11.

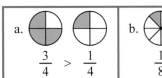 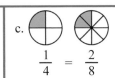

a. $\frac{3}{4} > \frac{1}{4}$ b. $\frac{1}{8} < \frac{1}{3}$ c. $\frac{1}{4} = \frac{2}{8}$ d. $\frac{3}{12} < \frac{3}{10}$

12. a. <   b. >   c. <   d. =

13. Since the two fractions have the same kinds of pieces (ninths), look at the amount of the pieces. Eight pieces is more than five pieces, so 8/9 is greater than 5/9.

14.

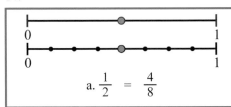

a. $\frac{1}{2} = \frac{4}{8}$ b. $\frac{1}{3} = \frac{2}{6}$

15.

 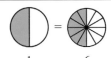 

a. $\frac{1}{3} = \frac{4}{12}$ b. $\frac{1}{2} = \frac{6}{12}$ c. $\frac{9}{12} = \frac{3}{4}$

16. No. She is not correct. The two wholes are not the same size.
If the wholes are made the same size, we can easily see
that 2/8 < 2/4.

Fractions Test, p. 71

1. a.   b. >   c. >   d. <

2. $\frac{1}{8} < \frac{1}{5} < \frac{1}{4} < \frac{1}{2}$

3.

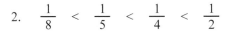

4.

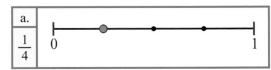

5.

   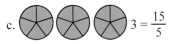

a. $1 = \frac{8}{8}$ b. $2 = \frac{12}{6}$ c. $3 = \frac{15}{5}$

111

6.

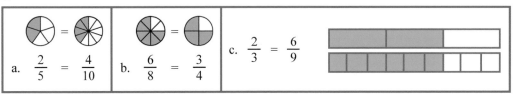

7. They eat the same amount of bread, because eating 3 pieces out of 12, and 2 pieces out of 8 signify the fractions 3/12 and 2/8, and they are equivalent fractions (both are in fact equal to 1/4).

8. It is wrong because the two wholes that we take the fractions from are not the same size. You cannot compare fractions unless the wholes they refer to are the same.

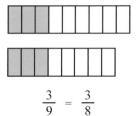

$$\frac{3}{9} = \frac{3}{8}$$

Mixed Review 17, p. 73

1.

a.	b.	c.	d.
$56 \div 7 = 8$	$48 \div 6 = 8$	$54 \div 9 = 6$	$48 \div 8 = 6$
$49 \div 7 = 7$	$72 \div 6 = 12$	$81 \div 9 = 9$	$72 \div 8 = 9$
$28 \div 7 = 4$	$54 \div 6 = 9$	$36 \div 9 = 4$	$32 \div 8 = 4$

2.

a. $7 \times 6 = 42$	b. $3 \times 0 = 0$	c. $9 \times 8 = 72$
$42 \div 7 = 6$	$0 \div 3 = 0$	$72 \div 9 = 8$
$42 \div 6 = 7$	~~$3 \div 0$~~ (not possible)	$72 \div 8 = 9$

3.

a.	b.	c.
$16 \div 5 = 3$ R1	$21 \div 4 = 5$ R1	$19 \div 6 = 3$ R1
$12 \div 5 = 2$ R2	$27 \div 4 = 6$ R3	$31 \div 6 = 5$ R1

4. Kathy needs to read nine pages of the book each day.

5. Each child will get three apples and there will be two apples left over.

6. $2 \times 3 + 4 \times 6 = 30$ OR $6 \times 3 + 4 \times 3 = 30$ OR $6 \times 6 - 2 \times 3 = 30$

7. Estimate: 140 ft + 80 ft + 140 ft + 80 ft = 440 ft.
   The real perimeter: 438 ft

8. a. 1/4   b. 2/7   c. 3/8   d. 6/7

9. Since the two fractions have the same amount of pieces, look at the size of the pieces. Eighths are larger pieces than tenths, so 7/8 is greater than 7/10.

10. a. 5,637   b. 6,121
    c. 9,696   d. 4,010

1.

a.	b.	c.
$7 \times 2 = 14$	$5 \times 7 = 35$	$7 \times 6 = 42$
$2 \times 7 = 14$	$7 \times 5 = 35$	$6 \times 7 = 42$
$14 \div 2 = 7$	$35 \div 5 = 7$	$42 \div 7 = 6$
$14 \div 7 = 2$	$35 \div 7 = 5$	$42 \div 6 = 7$

2. a. $4 \times 12 = 48$. She will need 48 apples.
   b. $36 \div 4 = 9$. Each piece is nine inches long.
   c. $1.77. Five quarters makes $1.25. Three dimes makes $0.30. In total, you have $1.25 + $0.30 + $0.22 = $1.77.
   d. Word problems will vary. For example: Twenty-four horses are arranged into six rows for a parade.
      How many horses are in each row? $24 \div 6 = 4$.

3. a. 5, 9   b. 35, 42   c. 12, 6   d. 28, 64

4. a. $7 \div 2 = 3$, R1; $9 \div 2 = 4$, R1
   b. $13 \div 5 = 2$, R3; $14 \div 5 = 2$, R4
   c. $21 \div 2 = 10$, R1; $47 \div 6 = 7$, R5

5. a. ounces (oz) or grams (g)
   b. pounds (lb) or kilograms (kg)
   c. feet (ft) or meters (m) or meters and centimeters
   d. usually feet (ft) or meters (m). Possibly miles (mi) or kilometers (km).

6. a. $5 \text{ m} \times 6 \text{ m} = 30 \text{ m}^2$
   b. 90 m^2
   c. 42 m

7.

a. $\dfrac{3}{5}$		b. $\dfrac{7}{8}$	
		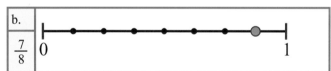	

Grading for Grade 3 End-of-the-Year Test

Instructions to the teacher: My suggestion for grading is below. The total is 207 points. A score of 166 points is 80%.

Grading on question 1 (the multiplication tables grid): There are 169 empty squares to fill in the table, and the completed table is worth 17 points. Count how many of the answers the student gets right, divide that by 10, and round to the nearest whole point. For example: a student gets 24 right. 24/10 = 2.4, which rounded becomes 2 points. Or, a student gets 85 right. 85/10 = 8.5, which rounds to 9 points.

Question	Max. points	Student score
Multiplication Tables and Basic Division Facts		
1	17 points	
2	16 points	
3	16 points	
	subtotal	/ 49
Addition, Subtraction, Word Problems		
4	6 points	
5	6 points	
6	4 points	
7	4 points	
8	4 points	
9	3 points	
10	3 points	
11	4 points	
	subtotal	/ 34
Multiplication and Related Concepts		
12	1 point	
13	1 point	
14	3 points	
15	3 points	
16	1 point	
17	2 points	
18	1 point	
	subtotal	/ 12
Time		
19	8 points	
20	3 points	
	subtotal	/ 11

Question	Max. points	Student score
Graphs		
21a	1 point	
21b	1 point	
21c	1 point	
21d	2 points	
	subtotal	/ 5
Money		
22a	1 point	
22b	2 points	
22c	2 points	
23	2 points	
24	3 points	
	subtotal	/ 10
Place Value and Rounding		
25	2 points	
26	5 points	
27	4 points	
28	2 points	
29	8 points	
	subtotal	/ 21
Geometry		
30	5 points	
31	2 points	
32	4 points	
33	2 points	
34	2 points	
35	3 points	
	subtotal	/ 18

Question	Max. points	Student score
Measuring		
36	2 points	
37	2 points	
38	2 points	
39	6 points	
	subtotal	/ 12
Division and Related Concepts		
40	2 points	
41	6 points	
42	3 points	
43	2 points	
44	2 points	
	subtotal	/ 15
Fractions		
45	6 points	
46	3 points	
47	2 points	
48	3 points	
49	4 points	
50	2 points	
	subtotal	/ 20
	TOTAL	**/ 207**

Grade 3 End-of-the-Year Test Answer Key

1.

×	0	1	2	3	4	5	6	7	8	9	10	11	12
0	0	0	0	0	0	0	0	0	0	0	0	0	0
1	0	1	2	3	4	5	6	7	8	9	10	11	12
2	0	2	4	6	8	10	12	14	16	18	20	22	24
3	0	3	6	9	12	15	18	21	24	27	30	33	36
4	0	4	8	12	16	20	24	28	32	36	40	44	48
5	0	5	10	15	20	25	30	35	40	45	50	55	60
6	0	6	12	18	24	30	36	42	48	54	60	66	72
7	0	7	14	21	28	35	42	49	56	63	70	77	84
8	0	8	16	24	32	40	48	56	64	72	80	88	96
9	0	9	18	27	36	45	54	63	72	81	90	99	108
10	0	10	20	30	40	50	60	70	80	90	100	110	120
11	0	11	22	33	44	55	66	77	88	99	110	121	132
12	0	12	24	36	48	60	72	84	96	108	120	132	144

2. a. 14, 24, 25, 36   b. 28, 40, 27, 35   c. 9, 16, 49, 32   d. 56, 30, 48, 54

3. a. 7, 5, 8, 7   b. 8, 5, 11, 7   c. 9, 7, 4, 9   d. 10, 8, 3, 3

4. a. 310, 149   b. 620, 344   c. 148, 80

5. a. 33, 5   b. 643, 45   c. 15, 378

6. a. 579. To check, add 579 + 383 = 962 using the grid.
   b. 2,476. To check, add 2,476 + 4,526 = 7,002 using the grid.

7. a. 7,153
   b. 792. Note that according to the order of operations the subtraction is done first.

8. a. △ is 294. Solve by subtracting 708 − 414.
   b. △ is 824. Solve by adding 485 + 339.

9. $83

10. 160 miles. Note that the half-way point is at 150 miles. They stopped at 140 miles (10 miles before 150 miles).

11. a. 800 light bulbs
    b. 736 are left. Solve by subtracting 800 − 64.

12.

13. 5 × 25 = 125. You can solve it by adding repeatedly: 25 + 25 + 25 + 25 + 25 = 125

14. a. 48   b. 20   c. 41

15. a. 7 × 4 = 28 legs
    b. 5 × 2 = 10 legs
    c. 8 × 4 + 6 × 2 = 44 legs

16. 8 tables, because 8 × 4 = 32, which is more than 31. Seven tables is not enough.

pagebreak

17. 3 × $8 + 3 × $6 = $42

18. She needs 7 bags. (Because 7 × 4 = 28.)

19.

	a. 10:51	b. 2:34	c. 3:57	d. 5:38
10 min. later	11:01	2:44	4:07	5:48

20. a. 45 minutes   b. 3:50 PM   c. May 28

21. a. 28 hours   b. 12 hours   c. 9 hours more   d. 48 hours

22. a. $25.54   b. $9.10   c. $12.70

23. a. $2.90   b. $0.55

24. $0.60. (You can add $2.35 + $2.35 + $2.35 + $2.35 = $9.40 to find the total cost.)

25. a. 700   b. 2,000

26. a. >   b. <   c. <   d. >   e. >

27. a. 5,700; 8,600
   b. 1,200; 7,800

28. a. 740   b. 990   c. 250   d. 670

29.

a. Round the numbers, then add: 3,782 + 2,255 ↓ ↓ 3,800 + 2,300 = 6,100	Calculate exactly: $\begin{array}{r} 3\ 7\ 8\ 2 \\ +\ 2\ 2\ 5\ 5 \\ \hline 6\ 0\ 3\ 7 \end{array}$
b. Round the numbers, then subtract: 8,149 – 888 ↓ ↓ 8,100 – 900 = 7,200	Calculate exactly: $\begin{array}{r} 8\ 1\ 4\ 9 \\ -\ \ \ 8\ 8\ 8 \\ \hline 7\ 2\ 6\ 1 \end{array}$

30. A - rectangle   B - square   C - rhombus   D - rhombus   G - rhombus
   Also, F is a parallelogram; however that is not studied in third grade.

31. Perimeter 22 units   Area 24 square units or squares
   Note that the student should also give the "units" and "square units" or "squares", not just a plain number.

32. a. Part 1: 108 m^2 Part 2: 270 m^2 b. 96 m
   Note that the student should also give the units "m^2" and "m" in his/her answer, not just plain numbers.

33. 9 inches.

34. a. The sides of the rectangle could be 5 and 3 units or 15 and 1 unit.
   See some examples on the right:

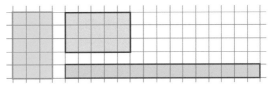

   b. The sides of the rectangle could be 1 and 4 units, or 2 and 3 units.
   See the illustration on the right.

35. $4 \times (2 + 5) = \boxed{4 \times 2} + \boxed{4 \times 5} = 28$ squares (or square units)

36. Check the student's answers.

   a.

   b. ████████████████████

37. mm   cm   m   km

38. ounces (oz) and milliliters (ml)

39. a. The mountain is 20,000 _ft_ high.   b. The pencil is 14 _cm_ long.
   c. Jeremy bought 5 _kg or lb_ of potatoes.   d. The large glass holds 3 _C or cups_ of liquid.
   e. The teacher weighs 68 _kg_ .   f. The room was 20 _ft_ wide.

40.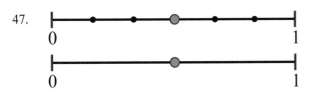

   $3 \times 6 = 18$ $18 \div 3 = 6$

   $6 \times 3 = 18$ $18 \div 6 = 3$

41. a. 17, not possible   b. 1, not possible   c. 1, 0

42. a. 8 R1   b. 4 R4   c. 6 R5

43. Can he divide the children equally into teams of 5? **No.**
   Teams of 6? **Yes.** Teams of 7? **No.**

44. Each child paid $10.00.

45. a. $\dfrac{3}{8}$ b. $\dfrac{7}{9}$ c. $\dfrac{2}{4}$ d. $2\dfrac{2}{5}$ e. $\dfrac{2}{3}$ f. $\dfrac{9}{10}$

46. 1 = 10/10   b. 2 = 10/5   c. 4 = 24/6

47.

48.

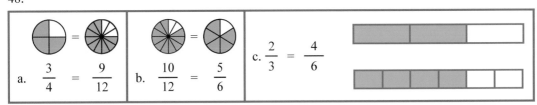

a. $\dfrac{3}{4} = \dfrac{9}{12}$ b. $\dfrac{10}{12} = \dfrac{5}{6}$ c. $\dfrac{2}{3} = \dfrac{4}{6}$

49. a. <   b. <   c. <   d. >

50. We cannot tell who ate more pie, because the two pies are of different sizes and it is not totally clear from the pictures
   which is more pie. And, even though the fraction 7/12 is more than 1/2, this thinking cannot be used here when the
   wholes are of different sizes.

Made in United States
Troutdale, OR
07/02/2023

10938947R00071